RELAUNCHING VIDEOTEX

RELAUNCHING VIDEOTEX

Edited by

Harry Bouwman

University of Amsterdam,
Department of Communication Science,
Amsterdam, The Netherlands

and

Mads Christoffersen

Technical University of Denmark,
Institute of Social Sciences,
Lyngby, Denmark

SPRINGER-SCIENCE+BUSINESS MEDIA, B.V.

ISBN 978-0-7923-1711-1   ISBN 978-94-011-2520-8 (eBook)
DOI 10.1007/978-94-011-2520-8

Printed on acid-free paper

CONTENTS

vi

FOREWORD

Helena Flam
Universität Konstanz
B.D.R.

Volker Schneider
Max Planck Institute for Social Research
Köln, B.D.R.

I.

A traditional sociologist or political scientist may find the choise of videotex as the object of this cross-national comparison surprising. Indeed, contemporary Sociology and Political Science have shied away from the studies of technology. Consequently, until recently they have not contributed much to the understanding of technological change, leaving this field of study to geographers and historians. The very best among such studies reveal, however, that the evolution of technology is a social construction and that the development and deployment of technical systems are intermeshed with social, economic and political relations (Hughes 1982). These studies show that technologies are often the result of the interaction between a number of social groups and actors (e.g. business, the state, etc.) as well as of social struggles revolving around the impact of such collaboration on the third parties. Once revealed, these complex interdependencies and processes are a compelling justification for the recent focus of sociologists and political scientists on technology and complex technical systems (Bijker et al. 1987, Burns and Flam 1987: 292-365; Mayntz and Hughes 1988). The aim of these as well as the present study is to uncover these webs and processes in order to contribute to the general understanding of technological change from a societal perspective. It is also to show that these processes are non-deterministic, interactive, and open-ended.

The major premise behind this comparative study is that videotex is not just a collection of different technological objects and techniques, but a development niche within which social agents cooperate and struggle around a variety of social, economic, political and technical issues. These interacting agents aim at the realization of their individual and organizational interests with the help of supportive institutions and societal rule regimes. In pursuit of their often conflicting interests and ideas, they attempt to reactivate old but also create new rule regimes and institutional arrangements. Not to be forgotten, these agents often create new organizations to spur on the process of technological change. Technical systems, in their turn, both in part resist and in part promote such restructuring of social relations. The core focus in this comparative study of videotex systems is this dynamic interdependence between the agents and the artefacts they create.

At a lower abstraction level, a technical system, here exemplified by videotex, may be studied from several different academic perspectives. What to a historian is a process of technological invention, innovation and diffusion, to a political scientist appears as a question of policy formulation, decision-making and implementation. Again, what to a political scientist is interesting as a process of interest-exclusion and -mediation, a sociologist may find an intriguing case of social mobilization pro and contra the new technology. Any evolving national videotex system can be said to go through the phases of technical invention, innovation and diffusion. But what makes for its distinctiveness is the particular mixture of private and state interests involved in its development as well as the specific situational constellation of the interests objecting to its specific aspects or proposed developmental trajectories. For it is these mixtures and constellations which reduce the number of the proposed options and, ultimately, give an artefact its context shaped design and trajectory.

1

H. Bouwman and M. Christoffersen (eds.), Relaunching Videotex, 1–5.
© 1992 *Kluwer Academic Publishers.*

II.

Like the technological system under scrutiny this book is also the result of a complex and "non-deterministic" process of "social construction" and interaction. A chain of "contingent" events contributed to this collection of case- and comparative analyses of videotex systems in a number of European countries. The idea of taking videotex systems as research objects for social scientists stems from Professor Renate Mayntz. As the director of the newly established Max Planck Institut für Gesellschaftsforschung, she set up a complete research program devoted to studying large technical systems from a sociological and political science perspective. She also initiated a comparative effort which issued in a joint project carried out with French and British colleagues between 1986 and 1990. Some of the results of the three national case studies and the cross-national comparison appeared in two books and several articles[1]. An integrated presentation will appear in the MIT Press' series "Inside Technology."

When in 1987/1989 Helena Flam - a Swedish-based scholar - visited the Max Planck Institute in Cologne as a Research Fellow, Volker Schneider and his British and French colleagues, Graham Thomas and Thierry Vedel, were already drawing their comparative project to its conclusion. At the time, Joanna Rose, became interested in the Swedish videotex at the Technical Museum in Stockholm. The idea of comparing the videotex development in smaller European countries with that in the larger ones emerged. In particular, Volker Schneider suggested, it would be useful to take a point of departure in Peter Katzenstein's theorizing (Katzenstein 1985). Katzenstein's research focused on the differences in the elite composition, decision-making structures and interaction styles of "small-democratic" versus "large-democratic" states. In their common research application[2], Helena Flam and Joanna Rose proposed to investigate the impact of the Swedish decision-making structures and styles on the development of videotex in Sweden. They also applied for funds to integrate also other small countries into the comparative effort whose ultimate aim was to test the Katzenstein-theses.

Using the project funds granted by The Bank of Sweden Tercentenary Foundation, the first comparative workshop was held in Stockholm on May 18-19, 1990. Tomas Ohlin, a Swedish researcher involved in the development of the Swedish videotex both as an observer and practitioner, contributed to the organization of this workshop by securing the support of Stockholm's Business School on whose premises the workshop participants met. The workshop included representatives of small democracies coming from Austria, Denmark and the Netherlands. Its primary focus was on the case-studies. The first attempts were also made to identify broad commonalities and differences in videotex development in a comparative perspective as well as to identify new trends.

During the Stockholm workshop, the initial intention of meeting again with the purpose of including even other researchers doing research on videotex in small democracies was stated. Preliminary publication plans were also formulated.

About six month later, in December 1990, Harry Bouwman relying on the funds of the Foundation 'Het PersInstituut' and Videotex Netherland organized the next workshop in Amsterdam. In this workshop, a representative of Belgium joined the group. During its course, it was possible to go beyond the initial emphasis on the case-studies and to start a search for a common framework. The idea now was that if all the case-studies were rewritten within this framework, a new set of comparative research questions could be addressed. In particular, several researchers wanted to use the case-study material to address the issue of what preconditions had to be met in order for a successful videotex market to develop. The concept of "critical mass" with which several of the present researchers already worked provided a label

---

[1] Mayntz/ Schneider 1988, Thomas/Miles 1989, Schneider 1989, Vedel 1989, Schneider et al. 1991.

[2] Bernhard Joerges contributed to the first research proposal.

for this problem area.

Mads Christoffersen organized the final workshop in Copenhagen. Funded partially by the Danish Teletechnical Society, Tele Danmark, Copenhagen Telephone Company, Jutland Telephone Company and Kommunedata. It took place in August, 1991. In preparation for this workshop, Mads Christoffersen and Harry Bouwman provided all workshop participants with a detailed chapter guideline. Thanks to their efforts, the representatives of Switzerland, Greece, Italy, Spain and the United States joined the original group and thus considerably extended the range of comparison. In Fall 1991, intensive final editorial and case-study work became completed.

III.

The result of this endeavour is an impressively broad and coherent cross-national comparison of videotex systems. The advantage of extending the comparison up to more than a dozen countries in the world while using the method of "comparative case studies" is that it enables one to go beyond the narrow focus on intentions, resources, actions and interactions of individual policy actors and technology developers which often predominate in single case studies. The broad comparative perspective used in this book facilitates the discovery of contextual variables and structural effects such as, for instance, national policy making styles, specific political and administrative regimes or market structures, which are invisible in a treatment of one country alone or a comparison of only two or three countries.

The "comparative-case-studies"-method also allows for the rapprochement between two distinct research perspectives: the history of technology vs. the inter-organizational decision- or policy making. In analyzing the introduction and diffusion of videotex in their respective countries, the contributors from a synchronical perspective examine the way in which the interplay of economic, technical and political forces affected the major decision-makers. From a diachronical perspective they also show how the technological decision-makers reduce the set of technical possibilities at one point in time to specific national 'trajectories' of technological and politico-economic development in subsequent phases. Their decisions get concretized in specific, national divisions of labour and authority and in particular technical solutions, applications and services.

The development of videotex shows interesting features from a synchronical as well as from a diachronical perspective. The whole idea started in Britain but quickly "sprang" over to France and Germany. The British, French and German videotex developers then served as technological leaders. The British and German videotex concept centered around a TV-set, equipped with a decoder. The French concept, in contrast, centered on a dedicated compact terminal. While in France the state distributed terminals free of charge and thus made them accessible to the broad public, in Great Britain and Germany videotex equipment had to be purchased. While the British Prestel concept was centralized, so that information had to go through a central computer (later, a computer set), the French Minitel was decentralized - from any terminal you can speak with a large number of other database computers. The German videotex developers at first adopted the Prestel concept, but then tried to decentralize it a la Minitel without abandoning its hierarchical storage structure.

Finally, the three leaders also differed in their marketing strategies. The 'direct' French approach contrasted with the German "sequencing" and the British 'management of expectations' approach. An entire range of countries tried to mimic their respective 'significant others.'To offer just a couple of examples, Austria oriented itself to the German, Sweden to the British and Belgium to the French concept. Interestingly enough, Sweden seemed the only country to draw some temporary benefits from its 'late-bloomer' status. For example, while it copied the technical Prestel concept, it very quickly oriented itself to a business-based rather than a household-based market, thus securing initial success for videotex in Sweden.

Contrary to our initial expectations, it turned out that the initial technological trajectories did not necessarily remain fixed. In reaction to more or less marked market failures, several national decision-makers narrowed or broadened their ranks and redefined the division of labour and authority. Alternatively, their composition did not change but they redefined their videotex concepts and tied together new specific technological solutions, technology users and marketing strategies. In several cases, a switch from a centralized Prestel to a decentralized Minitel videotex concept took place. Even BTX in Germany is gradually becoming more decentralized. Shifts between and combinations of the business-oriented and the public-oriented markets have also occurred or are in the planning stage. The most pluralistic seems the Dutch technological and political development which led to the emergence of a united 'hybrid' videotex system change into an open system, relying on a number of videotex technologies, publics and marketing strategies.

These reorientations make it very difficult to speak of strong 'path dependencies' in case of videotex (David 1985). Although politico-economical decisions and technical choices may create restrictions for further development, in a longer perspective videotex development is never completely locked-in into a given trajectory. While this might have been the case with some technological standards (MS-DOS, VCR) (David 1985, Arthur 1990), in the case of videotex it does not seem that the initial technological choices and expectations surrounding them have created such strong "lock-in" effects that an initially decided upon path was completely impossible to leave. The case-studies provide plenty of support for the claim that small historical events and contingencies may produce large effects. They also show that actors are learning and undertake reorientations even when this is costly. Technological development seems thus much more open-ended than a deterministic approach often suggests.

References

Arthur, B. (1990): Positive Feedbacks in the Economy. In: **Scientific American**, February 1990, 80-85.

Bijker, W. E./ T.J. Pinch/ Thomas P.Hughes (eds.) (1987), **The social construction of technological systems**. New directions in the sociology and history of technology. Cambridge: M.I.T. Press

Burns, T. / H. Flam, (1987): **The Shaping of Social Organization**. London: Sage.

Callon/ H. Law/ A. Rip (Hrsg.), (1986): **Mapping the Dynamics of Science and Technology. Sociology of Science in the Real World**. London: Sage.

David, P.A., (1985): Clio and the economics of QWERTY. In: **American Economic Review** 75, 332-337.

Hughes, T. P., (1983): **Networks of Power. Electrification in Western Society 1880-1930**. Baltimore: Hopkins.

Katzenstein, P., (1985): **Small states in world markets. Industrial policy in Europe**. Ithaca: Cornell University Press.

Mayntz, R./ V. Schneider, (1988): The Dynamics of System Development in a comparative perspective. Interactive videotex in Germany, France and Britain. In: R. Mayntz/ T. P. Hughes, **The Development of Large Technical Systems**. Frankfurt/New York: Campus, 263-298.

Mayntz, R./ T. P. Hughes, **The Development of Large Technical Systems**. Frankfurt/New York: Campus.

Schneider, V., J-M. Charon, I. Miles, G. Thomas, T. Vedel, (1991): The Dynamics of Videotex Development in Britain, France and Germany: A Cross-national Comparison, in: **European Journal of Communication** 6, 187-212.

Schneider, V., 1989: **Technikentwicklung zwischen Politik und Markt: Der Fall Bildschirm-text**. Frankfurt a.M.: Campus.

Thomas, G./ I. Miles, (1989): **Telematics in Transition**. London: Longman.

Vedel, T. (1989) 'Télématique et configurations d'acteurs: une perspective européenne', Technologies de l'Information et Société - **Réseaux** 2(1): 15 -32.

CHAPTER 1

INTRODUCTION. VIDEOTEX: IS THERE A LIFE AFTER DEATH?

Harry Bouwman
Department of Communication
University of Amsterdam

Mads Christoffersen
Institute of Social Sciences
Technical University of Denmark

Tomas Ohlin
Teleguide
Sweden

In this book an overview is given of the introduction of videotex in different European countries and the United States. Not much has been published so far about organizational and socio-political perspectives on videotex. We therefore approach this field with caution. Still, we hope to rise interest in enlarged circles about the characteristics of this important new type of communication.

In the history of technology during the 20th century we often find adoption patterns that resemble each other, although quite different types of technology are concerned. For television, for video, for facsimile, certain comparable time slots seem to have been needed for obtaining "market acceptance". Consumers need time to adopt a new technology - a time span often longer than expected. This indicates that there is a clear difference between **social** and **technological maturity** - a distinction very rarely made by innovators and engineers who are optimistic and eager to implement their new "gadgets".

Such time slots of multiple decades have been noted, the telephone took about 90 years to gain general coverage in the households. Looking at videotex, and discussing its "life after death", it seems reasonable to ask if time slots for adoption tend to decrease with the higher pace of technological innovation. Videotex is based on computer communication on existing networks, relying on humanly and friendly programmed forms for interaction between users, and users and databases. One of the most typical characteristics of computers is their flexibility, their possibility (at least in theory) to be easily programmed for different types of usage. Because of this flexibility - the possibility to be adopted to changing needs - the technological aspects of videotex may be accepted by users faster than other (and less computerized) types of technology. Still, the social factors surely need their time. Whether this sums up to a faster adoption of videotex compared to other "innovations" from a historical view-point, remains to be seen.

The aim of the book is to offer an analysis of the role major actors played in the telecommunications policy field regarding certain Value Added Services and of the manner in which telecommunication companies, national governments, information and service providers, hard and software industry tried to achieve their particular goals.

Fundamentally, we conceive of videotex as an **interlocked innovation**. This means that videotex is a combination of innovations on at least three levels:
- innovation in the telecommunication infrastructure;
- innovation in the supply of new services;
- social innovation in the way users fulfil their specific communication and information needs.

In order to be able to introduce a videotex system with success, three conditions have to be fulfilled: there must be an appropriate infrastructure, a kind of cooperation between system

7

H. Bouwman and M. Christoffersen (eds.), Relaunching Videotex, 7–13.
© 1992 *Kluwer Academic Publishers.*

provider, service providers and soft- and hardware industry has to be established and thirdly an articulated user demand has to be expressed.

The introduction of this 'new' type of telecommunication service is policy and technology driven. As such this introduction is interesting because at the starting time, there is no clear demand as such from the user side, neither from professional nor from residential users.

From the early days the idea existed that videotex was **the** consumer oriented technology of the coming information society (Toffler, Naisbitt, Mosco). In almost every European country the public discourse was articulated that acceptance of services from an easy-to-use terminal would be general. The world would be at the fingertips of every citizen, using this new device which was a combination of two old technologies: telephone and television.

But reality turned out to be more robust than anticipated. Videotex of the 1970s and 1980s showed to be a failure. Only in France the forecasts were met and the ambitious expectations were fulfilled. It is clear that when videotex was introduced, a lot of unforeseen problems had to be solved. Although the problems in general were different for each country, certain problems were common to all.

It is important to keep in mind that when the first ideas regarding videotex were presented the situation in the field of telecommunication, computer hardware and software differed considerably from the present situation. Personal computers and modems were not yet available. The hardware industry was attempting hard to develop cheap decoders for television sets to 'receive' videotex. The telecommunication networks were being gradually upgraded. Packet switched networks were not yet available. The first was introduced in Spain at the start of the eighties. They became part of the general telecommunication infrastructure in that decade. Electronic publishing was still in its infancy. Users, both professional and non-professional, had little if any experience with computers, information retrieval or datacommunication. The possibilities of telematics were still unclear both to the suppliers, the users as well as to the vendors of telecommunication and information technology.

We will look both at the regularities and the differences. Some countries were innovators while others lagged behind. The first experiments with Prestel in the UK took place in 1973, Télétel was introduced in France in 1982, Ireland will start with the introduction of a Télétel-like system in 1991. The regularities are due to the collective learning process. But the possibility exists that some countries did learn from the experiences of other countries.

1. The regularities

Two dominant scenarios for the introduction of videotex can be distinguished. Prestel and Télétel can lend their names to these scenarios. The scenarios differ on the following points:
- terminal design and strategy of terminal provision
- system architecture
- administrative system (including billing)
- marketing strategy
- regulatory constraints and political support.

Before dealing with these scenarios in more detail, it must be stressed that both Prestel and Télétel or any system like them could only develop within a specific policy and media environment (Schneider, Charon, Miles, Thomas and Vedel, 1991).

1.1 The policy environment

The motives of the main actors involved were often quite similar. The PTTs wanted to generate more traffic (and thus revenue) on their existing telecommunication networks and develop new services. They had to prepare for competition with the main computer and time-sharing companies, who were involved in developments of more or less digitalized networks. The governments saw a possibility to support the domestic telecommunication and/or consumer electronic industry. The interest of hard- and software companies speaks for itself. In some countries the (newspaper and magazine) publishers participated partly for defensive purposes: videotex was for a long time considered a threat to traditional publishing. In other countries publishers saw videotex as a possibility for conquering new markets. The last category was formed by other potential service providers, who saw possibilities to expand their interests: For instance publishers of the Yellow Pages, mail-order-companies, banks, insurance companies and so on. In other countries the broadcasting organizations or their representatives initially played a role as well. But with the crystallization of the difference between broadcast videotex (teletex) and interactive videotex, their role diminished.

The central players in the policy arena were the PTTs and the governments. They sometimes cooperated with other actors while at other times for reasons of interest priority they chose to block actively or passively the developments in the field of videotex. Other actors, often service providers such as the press, had a hard time conquering a position in this arena. Still others were hardly involved, most notably representatives of both the professional users and consumer organizations.

With the trend of deregulation in the 1980s the position of most PTTs did change considerably. Originally the PTTs were a branch of the central governments, but with the changing political climate their position became more market-oriented and they were forced to cooperate with various actors. These actors were sometimes dependent on the PTTs, but were also their competitors in other areas of the telecommunication market. The developments in the field of value-added services as for example in the UK and the Netherlands, are notable examples.
It is clear that the deregulation of the position of the PTTs differs in detail from country to country and is highly dependent on the policy climate of each country. The most liberal and deregulated situation can be found in the United Kingdom and The Netherlands. The most regulated position can be found in Germany, France, Italy and Belgium. The role of the EEC could also be considered in this context. The message of liberalization in the Green book of the Commission influenced the policy in the different countries in different ways.

1.2. The Media environment

Not only the policy environment and the legislation differ from one country to the other. The media environment in which videotex must position itself also has to be seriously considered. The success of Télétel and the failure of videotex in other countries can to a large extent be explained by looking at the media environment at the moment of introduction in the different countries. Information provision in France was low key at the moment of introduction of Télétel. Train schedules and phone numbers for example were hardly available by telephone. Voice-based information services were limited. Broadcast viewdata is not available in France, while this is heavily used in Britain, Germany and other countries. The penetration of home computers and modems on the other hand is lagging in France compared with the situation in other countries. These are some indications that should be taken in to account when comparing the two dominant scenarios regarding the introduction of videotex.

1.3 The Prestel scenario

Initially this was the scenario that was copied by most countries for example Germany, Denmark, Italy, Austria and Switzerland. The Prestel concept was based on a presumed adoption of videotex by the residential market. The consumer had to buy a television-set with a special videotex decoder (or a separate decoder) in order to acquire access via the telephone network to a central database in which information pages .were stored. The hardware for the consumer was expensive and not easily available at the moment that Prestel or for example the Danish or Austrian (the Prestel-standard mupid-terminal) videotex system became operational. The user had to subscribe to Prestel and prestel-like systems. The billing system was complicated and largely unclear to the user.

The system architecture was very complex and made it difficult for the information providers to update their information. Although the PTTs which introduced the Prestel-like systems occupied a very central position in the introduction of videotex, they hardly paid attention to the quality and reliability of the information offered or the market which they wanted to serve. At the moment that the experimental phases ended and videotex was introduced on the market, marketing efforts declined further. It became increasingly clear in those different countries which followed the Prestel concept that the adoption of videotex services would go very slowly and the number of subscribers wouldn't meet the initial expectations.

Since videotex successively appeared to be more acceptable in the business community, British Telecom began to shift their marketing efforts accordingly to the needs of specific business segments. The same switch in marketing activities following initial trials with videotex occurred in Denmark and the Netherlands. In Austria the PTT opted for closed users group and in-house use of BTX. In Sweden the business market was from the beginning regarded as the chief videotex user area, based on practical and financial arguments, but also because of fears for competition from new general purpose mass media. The failure of Prestel also played a role in the choice for a niche-oriented marketing strategy. However this strategy wasn't successful either. Consequently the break-even point was not reached in the countries which followed the Prestel-concept. Videotex was becoming a failure.

By the mid 1980s the original enthusiasm in many countries had crumbled away. The optimistic predictions on the number of subscribers were replaced by down-to-earth ascertainment of the reservations of service providers and users. Videotex was more and more considered a "much-ado-about-nothing" matter (Godefroy Dang Nguyen and Erik Arnold, 1985 and Noll, 1985). Many observers did put it bluntly: Videotex was dead!

The above scenario, which is described in more detail in chapter 2, was almost entirely copied by all European countries, except France and Sweden. Almost identical decisions and subsequently almost identical errors were made as if nothing was to be learned from previous experiences.

1.4 The Télétel scenario

The French experience was radically different. The introduction of videotex was considered to be a part of industrial policy. The first step was the modernization of the telephone system. During the second phase videotex was proposed as an attempt to generate traffic on the Transpac packet-switched network and at the same time to help create an information service community. Because the dedicated Minitel-terminals were provided for free the number of users could be "planned" in advance. The kiosk-structure made the billing system transparent: the user knows beforehand what he or she has to pay.

The system architecture is very simple: the Minitel terminal is connected to the telephone network, which is connected to Transpac. This is like other systems. However, in France, databases of the information providers are directly connected to the packet switched network. This simple architecture, in which no central host-computer is necessary, turned out to be efficient for the most important applications, and made it possible to reach a critical mass of services on the supplier side in a short period of time. Marketing of the services was not solely

done by France Télécom but mainly by the information suppliers themselves. Because suppliers competed for the same market many promotional activities were developed.
This contrasted to activities developed by those PTTs which introduced Prestel and Prestel-like systems.

It is clear that the French scenario, mainly based on the decentralized structure and the free hand out of terminals, is more successful than the Prestel scenario when one considers the amount of users reached and the traffic generated within the telecommunication network. In this context it is interesting to note that for example in Switzerland, but also in Belgium and Italy, French Télétel has been and is a strong competitor to the local videotex system in the French speaking parts of these countries. Télétel spills over from France to neighboring regions and countries. But more important is the trend to copy the Télétel approach with regards to terminal distribution and kiosque-billing, and to relaunch videotex in those countries in which the Prestel approach wasn't that successful as for example is the case in Italy, Sweden and the Netherlands.
The prominence of the 'French model' has, however, not been totally unaffected by the numerous arguments about its high costs. The price for winning the position as the world leader within the videotex field may very well be payed for with substantial economic losses and deficits for France Télécom for a considerable amount of time.

2. The differences

Although there are two dominant scenarios, the adoption of videotex in each country also has its own specific characteristics. Sometimes this has to do with the specific policy climate, sometimes with regulatory, cultural or geographical factors.

Concerning the United Kingdom (see chapter 2) it is clear that developments in other countries have bypassed Prestel. Nevertheless England is an interesting example of the consequences of deregulation for the development of a market for certain Value Added Services. The concept of videotex in England however is blurred. At the one hand one can see that Prestel is some kind of a curse, at the other hand private videotex systems did emerge aiming at professional users communities.

The case of France makes it clear that the initial technological advantage of the United Kingdom certainly is not the only factor that is important for a successful introduction of a new technology. More important and perhaps decisive was the public discourse concerning a new technology. The French Nora-Minc report had a tremendous impact on the public opinion in France. Although in every country the technology push dominated, only in France this push was successfully supported by handing out free terminals. Demand was stimulated by strong service marketing and positive public climate but also by the interventions of the DGT (France Télécom), which had an unquestioned leadership in the field of innovation policy. The idea of an information society has become somewhat of a reality in France.

In Germany the public debate played an important, but less stimulating role. The discussions in the parliaments both of the Länder and the Federation made the introduction of BTX a political issue. The main issue was if videotex had to be considered as a form of mass media or a tele-communication service. The question was raised what the (negative) social and economic consequences were. In other countries (perhaps except Switzerland and Sweden) public discussions were not that strong. In most countries the dominant discourse of the information society was used one way or another to legitimize the introduction of videotex.

Another problem is illustrated if we compare the Italian (see Chapter 3) and Dutch (see Chapter 4) case. In Italy, but also in the Netherlands, initially the Prestel scenario was followed.

However, in Italy the development of videotex was stopped, when it became clear that the Prestel approach wasn't successful. In the Netherlands different actors, who fulfilled different roles, launched new projects, which were mainly blocked by competitors or the PTT. All the different initiatives from network operators, service and information suppliers 'had to be integrated' considering the small size of the market. This made a specific role, that of **system integrator**, necessary. This integration of the different initiatives is also reflected in the system configuration. In Italy there was only one dominating actor who after a long period of relative inactivity relaunched videotex, based on the Télétel concept. No bargaining or blocking of initiatives was involved, PTT being the most dominant actor.

Some of the smaller countries were heavily influenced by the developments in neighboring countries with comparable cultural backgrounds. For instance Austria (see Chapter 5) and Switzerland (see Chapter 6) followed the BTX-approach. The problem of Belgium (see Chapter 7) is complicated by the fact that it is a bi-cultural country. Belgium had the example of the Dutch hybrid system and the French Télétel-system. On the other hand Switzerland is faced with comparable multi-cultural problems but seems to profit from it: users are happy to retrieve information from databases with different lingual and cultural backgrounds.

Switzerland followed an introduction strategy by decreasing the price of terminals and the communication tariffs. This policy is possible through cross-subsidization within the PTT. In Switzerland the Information Providers, especially through the combined efforts of the banks, play an important role in the introduction of services. This contrasts very strongly to the lack of interest from information providers in Denmark (see Chapter 8). In Switzerland the PTT performs the roll of system integrator, this is not the case in Denmark where the regional telephone companies have concentrated their effort on the building of the technical infrastructure while not paying attention to the usability of services. The lack of interest of the information providers in Denmark is thus mainly due to the dominating technical approach that accompanied the introduction of videotex.

The strategy of a decreasing price for hardware is also followed by the Spanish videotex service (see Chapter 9). However the decrease in price there is made possible partly because EEC-subsidies from the STAR-program are reallocated to the service operator, Telefonica. In Spain the information providers are not only the central actors, but the user community is also actively participating on the policy arena.

The Swedish approach (see Chapter 10) chose the business market for the marketing of videotex. In Sweden, like elsewhere, one is now confronted with the problem of how to attract sufficiently large groups of consumers. Ireland (see Chapter 11) seems to copy this Swedish strategy by directing the marketing directly towards the business sector. Although in the United States (see Chapter 12) videotex as it is known in Europe hardly exists, information services directed at the business market are highly successful. It has a spin-off to the consumer market in Prodigy, GEnie and other consumer oriented value-added services.

On the first hand it seems that innovation of value-added services in the United States contradicts the concept of interlocked innovation. However, the Prodigy case can be interpreted in the same way. IBM is at the same time service operator, system provider, hard- and software provider. But more importantly, it illustrates very clearly that in each country all of these elements have to be available at the same time and even then success will not be certain. Specially, the point of system integration is of interest. A powerful actor, as for instance the DGT (France Télécom), the PTT in Italy, Videotex Netherland, or a group of cooperating actors such as the Banks and the Swiss PTT in Switzerland, the system provider and the Users Association in Spain, or the triumvirate behind Teleguide in Sweden, has to take the lead.

What all these examples show is that even if there are a lot of regularities with regard to the introduction of videotex in Europe, there also are a lot of differences between the countries.

By the mid 1980s videotex was declared dead: it was a technical mis-perception not matching the needs of the users. The development in many European countries seems to support this view. But as mentioned earlier: new technologies have their, often considerable, adoption time. A series of new projects in countries as Spain, Ireland, Sweden make us ask the question if videotex in the 1990s is revitalised due to a shortening of the acceptance period - and not to its prematurely announced death.
The contents of this book may be seen in this perspective. Whether the acceptance time for future, and still more computerized types of technology will decrease further, is a matter for coming history to show.

References:

Naisbitt, J. (1982). **Megatrends. The New Directions Transforming our Lives**. New York: Warner Books.

Masco, V. (1982). Pushbuttons Phantasy. **Critical perspectives on Videotex and Information Technologies**. Norwood: Ablex.

Noll, A. (1985). Videotex: Anatomy of a Failure. **Information & Management**. Vol. 9 (99-102).

Toffler, A. (1980). **The Third Wawe.** Bantam Books Inc.

Schneider, V., Charon, J.M., Miles, I., Thomas, G., Vedel, T. (1991). The Dynamics of Videotex Development in Britain, France and Germany: A Crossnational Comparison. In: **European Journal of Communication**. SAGE, London, Newbury Park and New Delhi. Vol. 6, pp. 187-212.

CHAPTER 2

THE UNITED KINGDOM, FRANCE AND GERMANY: SETTING THE STAGE

Graham Thomas
Polytechnic of East London

Thierry Vedel
Centre d'Etude de la Vie Politique Française
Paris

Volker Schneider
Max Planck Institute for Social Research
Köln, B.D.R.

Introduction: The Problem of Disparity

So much has happened to videotex in Europe since it began in the early 1970s that it is becoming difficult to recall exactly what drove its development in the beginning, and what were the origins of the considerable variety in Europe's videotex landscape today. This chapter aims, briefly, to retrieve some of this knowledge through an examination of the origins and key features of the first European countries to attempt to introduce videotex on a grand scale: the United Kingdom, France and Germany.

It deals with the topic in less detail than the following chapters deal with developments in other European countries. Partly, this is because more has already been written about these countries than the others: there is therefore less that is new to say. Mainly, however, it is because the authors of this chapter have written at length elsewhere about the paths followed in the three countries[3]. Rather than simply refer the reader to this work, it was decided to present briefly its main conclusions in a way that is consistent with the aims of this book. In
particular, this chapter aims to draw out those features of videotex in the three countries which define their divergent paths and which 'mark out the territory' which has been explored - and extended - by the designers and promoters of videotex in other European countries.

The results of the three attempts to develop videotex are, as is well known, startlingly different. Although debates are still in progress on the topic of whether any of the flagship videotex systems in the three countries have made any money, or are ever likely to make money, for their operators, the disparity in scale of the systems today is clear. France télécom has installed a base of six million Minitel terminals, a figure which is still rising. Even if a considerable number of these stay unused "in the cupboard", as one commentator suggested, the number of services offered and the amount of traffic generated by the Télétel system are ample evidence of the successful implantation of videotex into the economy and society of France. In Germany, after a much-delayed start, Bildschirmtext has slowly built a solid-looking foundation of customers - around 300,000 at the last count. This is well below initial expectations but may prove to be a

---

[3] See Schneider et al, 'The Dynamics of Videotex Development in Britain, France and Germany: a Cross-national Comparison' in European Journal of Communication, vol 6, no 2, June 1991, pp 187-212. Also Schneider et al, Pathways to Telematics: the Politics of Videotex in Britain, France and the Federal Republic of Germany, forthcoming, MIT Press.

15

H. Bouwman and M. Christoffersen (eds.), Relaunching Videotex, 15–30.
© 1992 Kluwer Academic Publishers.

viable base on which to consolidate around a number of customer groups[4]. The country which began the whole story, however, has even less to shout about. Although the United Kingdom has shown that there is a healthy market for the provision of many kinds of telematic services, most of these are provided in the non-videotex formats. The situation is made complicated by the existence of competition at both the service and network levels - British telecom's Prestel enjoyed a monopoly of videotex services for only a short time - but even if all the videotex services which are available to subscription or use by third parties are added together, the number of videotex users falls far short of Germany's total. Prestel itself, having recently closed down its main service to non-business users, probably had fewer than 50,000 subscribers by the end of 1991.

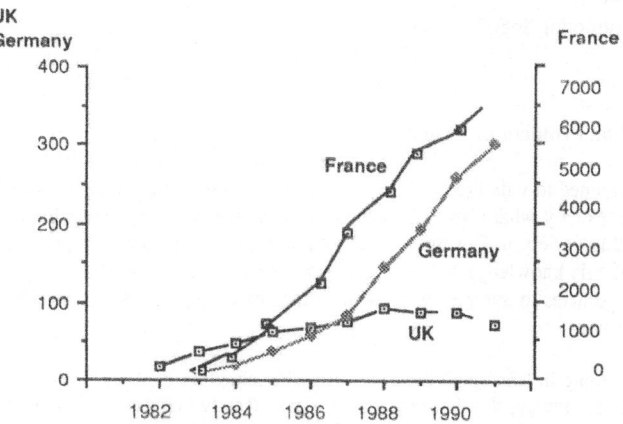

Figure 1: Videotex subscribers in Britain, France and Germany

So, why has there been this huge difference in scale? What were the key factors which explain the uneven development of videotex in the three countries. The following sections seek to lay them out in a form which, hopefully, also illuminates the choices made by videotex developers in the other countries covered in this volume. They will look in particular at strategies in technical design and marketing which aimed to create demand for a new 'product'; at the means through which the various actors involved tried to coordinate their efforts, to promote their aims and to forestall actual and potential opposition; and at the political and regulatory environments which sometimes opened up avenues of development and at other times constrained the directions in which development could take place. Before examining these themes in a cross-country comparison, however, the three stories will briefly be outlined.

The First One Now Shall Later Be Last: Pioneering and Privatisation in Britain

Although the origins of videotex can be traced to innumerable innovations in the fields of computing and telecommunications, and the idea of something like videotex had been current in most parts of the developed world since at least the end of the 1960s, it is generally agreed that videotex first took its modern shape in the United Kingdom. Specifically, videotex was developed

---

[4] The confounding of forecasts is perhaps the one truly common trend in the story of videotex around the world: even in France the forecasts of the Direction Générale de Télécommunications (DGT) were well in advance of the actual distribution curve..

by a team led by Sam Fedida at the UK Post Office's telecommunications Research Laboratories, firstly at Dollis Hill in London, then at new facilities at Martlesham Heath near Ipswich.

The British Post Office was indisputably the main actor at the beginning of videotex (or 'Viewdata', to give it the name initially favoured by the developers and still occasionally used today in the UK). Fedida's team laid down the foundations of the new service: it was to provide a large database of information, held on computers, which could be accessed by remote users via ordinary telephone lines, using simple terminals and a friendly user interface suited to non-experts. Other features were defined during a sometimes lively debate between technical and marketing people within the Post Office: the terminals should use colour to increase the friendliness of the interface, the service should be aimed at a mass-market and not just at professional and business users, a two-way messaging system should eventually complement the one-way retrieval of information, etc.

Gradually, the early visions became embodied in a service concept which it was deemed practical to launch. This reflected the chief goals of the Post Office. Firstly, Prestel (the name which replaced Viewdata for legal reasons) was to make money by generating extra telephone traffic at a time when the Post Office was nervously beginning to wonder when the market for telephony might become saturated. Secondly, money could be made from renting the facilities of the system (primarily the 'pages' on which information could be stored) to external information providers. Thirdly, the Post Office could be shown to be an innovative, go-ahead organisation at a time when state-run industries were coming under attack. And fourthly, the Post Office as a public service wanted to serve the public in new ways, providing a service which was both genuinely new and democratically accessible to the broad mass of people - and which also provided a boost to British industry.

Early on, the Post Office took the decision to bring other actors into the fold. Television manufacturers were to make specially-equipped TV sets which could act as Prestel terminals. The Prestel screen format would conform to that being developed by the television broadcasters for their teletext services. Consumer electronics retailers would sell the terminals and provide front-line advice to potential customers. The Post Office's biggest telephone switch supplier, GEC, would build the computers to hold and process the information. All manner of organisations would be allowed to become information providers on the system (indeed, the Post Office wanted to restrict its own information-providing role to the bare minimum, producing an index and general service information).

The reasons for this strategy of incorporation will be discussed later. But, following an inconclusive period of field trials, when the service was finally launched in late 1979 in the London area and nationally the following spring, the weaknesses of the strategy soon became apparent.

Firstly, there were delays. The TV manufacturers, unused to the production of sophisticated microelectronics at that time, did not have enough sets ready for the launch. Worse, those that were ready were very expensive, because the manufacturers saw Prestel as a high-margin venture which could offset the downward pressure on prices exerted in mainstream markets by Japanese and other competition. The retailers were largely untrained for their proposed role as advisers to the public. Information providers were still in the process of defining what kinds of service might be useful or potentially successful, and had trouble creating their electronic databases and transferring them, using primitive editing and uploading software, to the Prestel database. The Post Office's own counter staff and telephone engineers were not fully prepared to advise on and install the new service.

These difficulties and delays, understandably facing all the participants as they struggled to produce such a major innovation, were exacerbated by the lack of a central controlling presence. Perversely, the Post Office disbanded its Prestel marketing organisation around the time of the service's launch, believing that the terminal suppliers and information providers would take on the crucial marketing role. Potential customers were confused: there was no single place they could go to in order to find out about the Prestel service, order it and have it installed; and for many, there was no vital reason to subscribe - just a general promise of "information at your fingertips". Those with persistence, enthusiasm and the money for the equipment were faced with a database of uneven quality and indifferent organisation.

It soon became apparent that the projected subscriber figures were not going to be achieved, and less than two years after Prestel's launch British Telecom (as the Post Office's telecommunications business had become) re-evaluated Prestel's strategy and importance. Some regional computer centres were closed, and some Prestel staff transferred to other duties. More importantly, Prestel's facilities became oriented to those business sectors which had shown a proven need for its unique attributes - interactivity and rapid updating of information. In particular the travel trade became the mainstay of videotex (not just of Prestel, but of competing services from Istel, the Midland Bank and the Thomson travel organisation too), and niches were found for financial services, insurance, education and - the one exception to the business sector orientation - microcomputer enthusiasts.

In the following decade, British Telecom made a series of incremental improvements to the technical side of Prestel, notably the introduction of mailbox services and the connection of external computers to Prestel via gateways. Importantly, BT itself took control of most of the key information services, shrugging off the occasional argument with other information providers who felt they were being disadvantaged. This takeover ultimately proved fatal for the microcomputing services - BT closed down Micronet towards the end of 1991, after price rises and the ending of the popular 'chatline' services (on moral grounds) had led to the loss of over a third of its customers[5]. In 1988 a study was made to test the feasibility of a relaunch along the lines of Télétel. It was eventually decided not to attempt this, partly because of worries about moral outrage if 'messageries roses' were to come to the UK, but more importantly because BT did not think that a mass-market service would be profitable enough to risk the investment, and was worried that the mass distribution of terminals might lead users to access services (and even networks) run by other organisations.

Prestel never grew to anything remotely approaching the forecasts of its founders. The highest-claimed figure for 'terminals attached', before BT stopped publishing figures in 1988, was 95,000, and because businesses were assumed to 'attach' more than one terminal, actual subscriber figures (not published) were significantly less than this. It is unlikely that there were so many 'terminals attached' by the end of 1991, given that most of Micronet's 20,000 subscribers had been lost since 1988. Moreover, BT's 1985 claim that Prestel was running at an operating profit may not have remained valid during the recession of the late 1980s, so the prospects for the world's first videotex service at the beginning of 1992 were not good. It is possible that some services will be merged with non-videotex online services, and reorganised as part of BT Tymnet's international value-added service portfolio.

---

[5] The closure of the chatlines, which were but a pale reflection - technically and otherwise - of the French messageries, was partly a 'knock-on' effect of concern about the content of certain audiotex services.

The 'Plan Telematique': Implantation and Innovation in France

The project to introduce videotex in France can in one sense be seen as the continuation of French industrial policy in the post-war period, involving the use of technology to (re-)gain control over domestic markets in the face of external threats and to promote exports. In another sense, though, Télétel was a major departure, involving new actors (in particular the PTT Ministry, the Direction Générale de Télécommunications(DGT)) and a new approach to marketing.

The 'Plan Telematique', which was produced in 1978 partly in response to the celebrated Nora/Minc report on the computerisation of society, was both a natural successor to earlier initiatives (e.g. the 'plan calcul' in 1964 which aimed at revitalising the French computing industry) and an attempt to push forward into an area where France would not only follow existing trends but establish a comparative advantage in a new industrial field. The organisation chosen to implement the plan, the DGT, was at the time involved in a major effort to modernise French telecommunications, a project which was perceived to be very successful and which was accompanied by changes in the DGT's organisation which gave it more flexibility and autonomy and prepared it for a leading role in the new plan.

Work on several new telecommunications-based services, including videotex, had been going on for some time. By 1978 plans were sufficiently far advanced for videotex to be seen as the most promising candidate to carry forward the French telematics programme. In fact, the plan allowed for a number of other activities besides videotex (e.g. touch-tone telephones, teletex, telefax) but the plans for videotex already included key features of what was to become Télétel.

These features included: the free distribution of terminals made to the DGT's specification; the introduction by the DGT of a 'magnet service' which would justify this distribution - the electronic directory; the use of a standard, unified, packet-switched network to carry the services (Transpac started operation as early as 1978); and co-operation with other actors to allow the introduction of other services not provided by the DGT. The DGT would commission the terminal equipment, the Minitel, and set the network and display standards; it would be able to control the infrastructure while allowing external organisations to add value to the system by introducing their own services. Crucially, the DGT did not want to manage databases for other organisations or make money from the storage of information provided by others; from the start, then, it was prepared to allow cheap and flexible connection of external host computers to the network, and to encourage external service providers via favourable access and tariffing conditions.

The development of the programme proceeded on the basis of these key features, but modified by the interaction with other actors with a stake in the development (or against the development) of the new services. Because of its high-profile government backing, the formation of Télétel took place in the full glare of public scrutiny. The press in particular put up stiff opposition at first, but were placated by concessions on the (temporary) prevention of firms from other sectors from entering the provision of news services and by giving the press a monopoly on classified advertising. Also, to appease potentially powerful political opposition, principles were established in 1981 which gave regional authorities power of approval before the electronic directory service was started in their region, and which made use of the directory service voluntary (in contrast to the DGT's early plans which envisaged the complete withdrawal of paper directories).

A series of field trials between 1980 and 1983 allowed the testing of technologies, services and marketing concepts. Perhaps the most interesting result of these was the development of real-time interactive services involving two or more users (which became known as 'messageries'), after users discovered a side-effect of an application which allowed interaction with the technical manager of a service in Strasbourg. In February 1983 the Electronic Directory service was formally inaugurated, and the following year the 'kiosque' access/tariffing system for external services was introduced (more on this below) and the free distribution of terminals on a region-by-region basis commenced. Unlike in the UK or Germany, there were no significant delays in the production or delivery of the terminals.

Within a year of launch it became clear that Télétel was going to be a success. Easy access, the popularity of certain services (including the Electronic Directory and the messageries) and lots of free publicity from the press (which was especially fond of the social analysis of messageries with a sexual orientation, the 'messageries roses') all contributed to the formation of a large and active user base. Within five years there were well over 10,000 different services - i.e. service code names, several of which could be offered by the same service provider - available for access, and terminals had been distributed at the rate of around a million per year.

Changes since the start of the full service have included: the proliferation of new access numbers and service tariffs in order to adjust service prices to what specific markets can bear, at the cost of diluting the original simplicity of the kiosque idea; a rationalisation of the number of service providers as larger firms have increased their market share at the expense of smaller ones; a shift from recreational uses towards professional uses, aided by tighter regulation and higher taxation of the messageries roses; and a slowdown in the distribution of free terminals, partly in the face of doubts about the economic profitability of the Télétel venture. In addition, various incremental technical improvements have taken place including a facility to switch between services without having to back out to the Télétel 'front page', and an attempt has been made by France télécom (without, so far, great success) to start a general electronic mail service based on Télétel. As can be seen in other chapters, the Télétel concept is now being promoted more in the export market, and international access to Télétel has been greatly extended since the late 1980s.

While questions have been asked about the narrow economic return provided by Télétel to France télécom, there is no doubt that videotex has had a major impact in France in both economic and social terms. The revenues to hardware, software and service providers; the generation of new employment and the development of telematics-related skills; the benefits to users in terms of new services and new modes of access to old services: all must be taken into account when drawing up the final socio-economic balance of Télétel. Compared with the development of videotex in other European countries, any 'failure' can be seen only in relation to the original forecasts of the originators of French videotex. It is the sort of failure which other system operators might envy.

Caught in the Legal Web: Disputes and Delays in Germany

In many ways, the development of videotex in Germany has paralleled that in the UK, although modified by a very different regulatory climate and by the wisdom that comes from hindsight. Bildschirmtext began as a proposal to transfer the 'viewdata' concept, and the technology, from the UK in 1975, and was initially promoted by the Deutsche Bundespost (DBP) for many of the same reasons given by the UK Post Office: the overcoming of potential telephone saturation, the chance to show that a state-run enterprise could be dynamic, a boost to the national electronics industry and other parts of the economy, and the delivery of useful new services to the public.

The DBP actively worked on the concept from 1976. Unlike in the UK, however, and more like in France, the government positively promoted the idea, firstly by taking note of an approving statement from a 'Commission for the Development of the Telecommunications System' in

December 1975, then as a result of a direct recommendation from a special study group in 1977.

The original conception of the system closely followed the British model. The DBP would run the network, own the computers in which the data was to be stored, and define the overall structure of the database. The contents of the database would be furnished by independent information providers, and the terminals - modified television sets - would be provided by the television industry. As in the UK, the chief target users of the new videotex services were to be private households.

However, the regulations surrounding German videotex were very different from those governing the development of Prestel in Britain. For historical reasons, West Germany had tightly-defined rules about what sort of organisations should control the different media. Videotex, as a prime example of media convergence, immediately stirred up an intense political argument amongst the press, the broadcasters and the federal and state governments (the 'Bund' and the 'Länder', respectively).

In a 'tight' regulatory environment, control of videotex depended crucially on how the service was defined. If it was conceived as an individual telecommunications service, it would fall under the control of the federal government (which was also responsible for the DBP); if it was a broadcast medium, it would be controlled by the state governments, the Länder; and if it was primarily an 'electronic newspaper', no arm of government should run it, as the press was in private hands and was market-driven.

In the regulations which were eventually enacted, the press was accommodated by assurances that the information on Bildschirmtext (btx) would remain in private hands and that the DBP would not expand its information services beyond electronic forms of its existing directories. The press had also argued against the DBP owning the storage medium, and here there was a compromise (which also made good technical sense) allowing the additional connection of external, privately-owned computers to the system. On the matter of whether videotex was an individual or broadcast medium, an accommodation between Bund and Länder was eventually (in 1983) reached, which ceded most of the control to the federal government, effectively handing day-to-day control to the Post Office. Further regulations in 1983 enhanced users' data privacy and forced the DBP to change the interface to btx so that customers had to confirm they wanted to choose any page which carried a viewing charge. By the time it began service, btx was probably the most regulated public videotex service in the world.

Initially, the question of the definition of btx as individual or broadcast medium was to be decided after gathering evidence from the field trials. Large-scale trials were inaugurated in 1980 but took some time to become fully operational. Industry participation was strong, especially from the members of the politically important community of small and medium-sized businesses - who supported the DBP's ownership of the storage medium as a means to allow 'democratic' access to small companies which did not possess large data processing systems. Because entry to the trials was expensive (the result mainly of investment costs), participants wished the government to end the uncertainty over whether the full service would be allowed to begin. As a result of this and the promotion of videotex by the DBP and its allies in government circles, particularly in the economics ministry, a decision to proceed with btx was taken by the German cabinet in May 1981, before the final reports from the trials were written and before an agreement was reached on the competencies of Bund and Länder in relation to videotex regulation.

The date for the official introduction was set as autumn 1983 (a year later than the original expectations of the DBP). In fact, the effective introduction was delayed further for technical and commercial reasons. Following the cabinet decision, the DBP put the contract for the full system out to tender. The eventual winner (after a delay due to political intervention in favour of Britain's GEC) was IBM with an ambitious concept of a semi-decentralised system involving regional computers with a mixture of their own datasets and data fetched from the central computer. It was an open secret that IBM could not develop this system in time for the official launch date; however, the terminal manufacturers were not unhappy with the delay, because (as in the UK) they found that development of the decoder chips was a more difficult process than they had predicted. The decision of the DBP in May 1981 to use the complex CEPT 1 display standard had not eased their task.

Bildschirmtext was, in fact, formally inaugurated as planned in September 1983. However, a provisional solution involving the GEC equipment used in the field trials had to bridge the gap until IBM could bring up its system in June 1984. Network expansion, though, proceeded quickly, and 95 per cent of households had local call access to the system by December 1984. Following its launch, btx experienced many of the same problems as Prestel had four years earlier. Few domestic users subscribed, partly put off by the high costs of adapted TV sets (eventually, most terminals in use were either dedicated 'dumb' terminals or personal computers with appropriate communications hardware and software). Business users showed proportionately more interest, but numbers grew much more slowly than had been forecast. The total number of subscribers rose steadily but slowly to around 300,000 at the end of 1991.

There was a healthy initial growth of information providers, helped by temporary low tariffs which had been conceded partly to compensate for delays in launching the system and partly because small businesses had put considerable pressure on the DBP when its first tariff concept was announced. When full charges were introduced in Spring 1986 there were 4,000 IPs registered. This figure then dropped sharply, and a process of rationalisation slowly reduced numbers further, but by 1990 there were still over 3,000 IPs in existence, and the number of pages on the system, which had also dropped in 1986, had crept back up to 700,000. More information was accessible to users via external computers, which had increased steadily in number to around 300 in 1990.

Few services were immediately profitable, though. Btx quickly shifted its focus towards professional uses: in 1985 its new marketing strategy gave up the orientation towards the diffuse residential market and emphasised discrete target groups such as insurance and travel agents, journalists and self-employed professionals. Certain other changes were made in the light of experience elsewhere. Unlike British Telecom, the DBP expanded its marketing budget. While it did not follow the by then successful French model of free terminal distribution, it did attempt to 'kickstart' the growth in subscriber numbers by commissioning and then renting cheaply a significant number of terminals (initially 50,000 'Multitels', then a larger number of less sophisticated 'Bitel' terminals).

Whether Bildschirmtext is seen as a failure or a modest success depends on how the comparison is made. Set against the French success or the initial forecasts of millions of subscribers within a few years of launch, it is clearly a failure. However, compared with Prestel in the UK or a number of other systems in Europe, it has not done too badly. Its future is open: certain technical developments might help it, e.g. ISDN makes the operation of btx much faster and hence more attractive to use; but deregulation of value-added services will allow competing services into the market.

The Importance of System Design

The brief descriptions above already contain some implied interpretations of why the success rates of the three systems differed so greatly. But this question should be addressed more systematically: just why was it that the telecommunications organisations in the three countries, organisations with seemingly similar technical resources and at least partly overlapping intentions, should produce such varying results?

One level of explanation looks at the designs and structures of the three systems and their suitability for their intended markets. The following is a much simplified categorisation, but the key design characteristics of videotex in the UK, France and Germany can be described under the following headings:

A. THE ORGANISATION OF THE DATABASE

In Germany and Britain, videotex originally relied on a central database to be accessed by all users. With Prestel, the database is essentially replicated on a small number of computers and users are routed to one of these when they access the system. In Germany the central database is augmented by regional databases, and there are sophisticated algorithms to control storage and transfer of data between computers, but the DBP manages both the regional and the central databases.

In France, on the other hand, France télécom manages the addresses of and routes to the host computers, but the computers themselves are mostly owned by external organisations, the service providers. This gives service providers more control over their data and, within limits, allows them to organise their data in the ways they find most suitable. Thus, there is a greater degree of flexibility built into the French system.

These two models are in reality not quite so different, as in both Germany and the UK external organisations can connect their own external computers via gateways to the videotex access systems. However, in the UK gateways were distinctly an afterthought and both there and in Germany the costs and complexity of running a gateway are considerable. Both BT and DBP Telekom see gateway access as an alternative to IP organisations renting pages on their systems, and charge accordingly. France télécom is more concerned with increasing the traffic generated by the system, and so has an interest in making access for external organisations easier and cheaper.

What has been contrasted as the 'logic of storage' versus the 'logic of traffic' is a major defining point of the differing concepts of the videotex service in France on the one hand and Germany and the UK on the other. Given that one of the reasons often cited for the failure of Prestel is the lack of control the UK Post Office had over the service as a whole, it is interesting to note that the designers of the most successful system in Europe deliberately chose not to control the organisation of the databases run by information providers. This raises the question of which elements in a videotex system are the most important to control - a point covered below. Also, it should be borne in mind that the decision to base a videotex system around a large-scale, centralised database ties the system closely to that database architecture and makes subsequent changes more difficult[6].

---

[6] This is especially true where, as in Britain, the problem is compounded by a shortage of people with skills in programming for a now uncommon computer architecture.

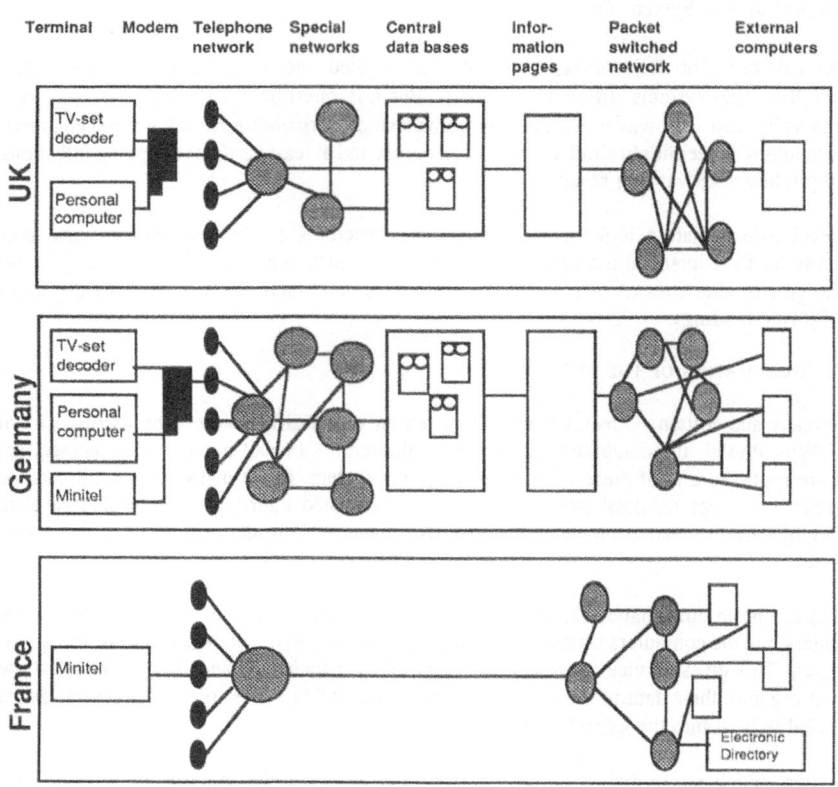

Figure 2. The Structure of Videotex in Germany, France and Britain

B. THE CHOICE AND DESIGN OF THE CARRIER NETWORK

The networks chosen to carry the videotex systems and services were partly a function of the service concept as described above and partly a consequence of timing. In France, Télétel was from the start designed to run on the national public packet-switched data network, conforming to international standards. This allowed host computers to be set up wherever there was a convenient access point to this network, and the network itself could be the beneficiary of economies of scale, as it could carry both videotex and non-videotex data traffic.

In the UK, Prestel was developed and launched before the full start of the public packet-switched network. In any case, the Post Office was using its own computers for the system and wished, perhaps, to keep its own networks distinct from those used by external organisations. Later on, when external computers were gatewayed to Prestel, the public data network was used for these connections, but the links between the Prestel 'information retrieval centres' (the host computers) and the user are made via the switched telephone service and a special Prestel access network. In Germany, users connect to the regional centres via the switched telephone service; a leased-line packet connection links the regional computers to the central database, and the public data network is used to connect external computers. Particularly in the British case, the need to maintain a dedicated access network for videotex has limited the extent of scale economies and the benefits of standardisation, and has slowed the pace of technical improvements (e.g. in access speed and incorporation of error correction).

C. TERMINAL DISTRIBUTION STRATEGY AND CHOICE OF TERMINAL

When the Post Offices in Germany and Britain began to develop their videotex systems in the mid-1970s, primarily for a household market, the decision to base the terminals around the household's television set made eminent sense. Home computers were then unknown, and most households already possessed (and were consequently familiar with the use of) the two basic technologies of TV set and telephone. The incorporation of the TV industry into the 'videotex coalition' was also seen to have the advantages of spreading investment costs, providing a ready-made distribution channel and making allies of potential opponents. The strategy failed for the reasons given in the description of prestel above. It should additionally be noted that the TV manufacturers insisted on supplying new sets for Prestel customers rather than discrete adapter boxes, and in Germany, too, the first terminals were expensive integrated TV/decoder combinations.

The terminal distribution strategy in France contrasted starkly with that of the other two countries. It was a strategy of 'implantation' based partly on the way in which telephone sets had traditionally been distributed - i.e. the PTT designed the terminal, commissioned manufacturers to produce it in large numbers, and controlled its distribution. The novel element was that distribution was free, as the Minitel was given to households as a replacement for the paper telephone directory. The cost of this distribution was legitimated on the grounds of internal rationalisation (savings in the cost of printed directories) and industrial policy (aiding the electronics industry).

The result of the different strategies was that, when the services were launched, households in Germany and Britain could not obtain inexpensive terminals, whereas households in France (or rather in the regions of France selected for distribution at the time of launch) had them handed out free. This was undoubtedly one of the most important reasons for the different rates of videotex usage in households.

D. ACCESS AND BILLING CONDITIONS

One of the other important reasons for the greater French usage was the ease of access to the videotex services and the associated transparency of the billing system. Interestingly, certain features in this area are the same for all three systems: for instance, users are billed by the system operator, which in turn transfers a portion of the money collected to the information providers. However, the differences are more important than the similarities.

In Britain and Germany, potential users must deliberately subscribe to the videotex system and pay a standing charge. The total charges they then pay are made up of a series of discrete elements: telephone costs (which appear on a separate bill); the standing charge; the cost of time spent using the system, which varies according to the time of day; the cost of sending messages to other users; and - for certain pages - the cost of viewing the page. This situation becomes even more complicated when closed user groups and certain other services (via external gateways) are taken into account, because service providers may add their own subscription and/or time-based usage charges. It can thus become difficult for user to know exactly how much a particular session will cost, and users can be faced with several distinct 'buying decisions' in the course of one access session.

In France, no subscription is required, and the way videotex services are tariffed is the same as that of premium-rate 'audiotex' services in most countries - a fixed unit charge is levied for calls to a specific access number. Originally in France there were just three access numbers (and hence three tariff levels) for videotex services. The number for the electronic directory allowed free calls for the first three minutes of each call, then the normal local telephone charge; a second number for 'in-house' services was cheap for the caller but enhanced by a payment from the service provider to the DGT; finally there was the 'kiosque' tariff for the vast majority of third-party services. The charges appeared on the normal telephone bill. The system was thus easy to understand and use, and charges were relatively easy to predict.

As mentioned above, the kiosque system in France was modified in 1987-8 in response to demands from service providers for a greater range of tariffs. Several more numbers, each with its own tariff level, were introduced, and so the original simplicity of the system has been somewhat compromised. However, the basic premise of one tariff for each access number remains, most services aimed at household users have remained at the basic kiosque tariff, and Télétel was already well established before the changes were made, so the changes have not had an unduly adverse effect on use levels.

Best Practice in Innovation, and Network-Based Systems

The design features described above help to illustrate how some of the lessons from the study of success and failure of innovations can be applied to the case of videotex. An obvious, but key, factor in successful innovation in a market economy is the delivery of a product for which there is sufficient market demand. Unless the innovator is exceedingly fortunate, it is unlikely that any significant product in the form in which it was originally conceived will be precisely tuned to meet market demand perfectly, and so products are normally altered in the course of development. The best way to ensure that any changes are in fact improvements is for the producer to monitor the responses and use patterns of users and to alter the products in accordance with users' wishes or perceived needs. This process may need to be continued throughout the lifetime of the product, and so the encouragement of user feedback and product redesign have become part of the 'best practice' of production, especially in sectors characterised by rapid technical change.

Of course, some types of product are more easily redesigned than others. Each of the videotex systems covered here was conceived as a large-scale network-based system requiring a considerable amount of initial investment, a fact which limits the amount of design flexibility. However, it can be argued that the French system, because of its widely decentralised database concept, was less restrictive than the other two. All countries made use of field trials, but in France the many different service providers, using a wide variety of host computers and database organisation, were able to adapt to the results of the trials and to the results of later mainstream usage better than the service providers in Germany and the UK who had to work within the constraints of a central database.

Another lesson learned by innovative firms is that they must ensure they have access to or control of a variety of 'complementary assets' necessary for the diffusion of the new product - typically such things as component parts, software, transport and distribution and marketing channels. Complementary assets for the providers of a videotex system might (depending on where the system boundary is drawn) include telecommunications networks, a terminal distribution system, and the production of services. When videotex was launched, each supplier controlled its own telecommunications network. As regards the production of services, the German and British Post Offices initially maintained a 'hands off' approach (although British Telecom was later to change its position) while the DGT supplied a key service - the electronic directory - while remaining a 'common carrier' for external suppliers of other (non-competitive) services.

Crucially, in France the DGT maintained close control of the terminal distribution system. This allowed it to overcome the problem of reaching 'critical mass' in a direct manner. For services based on telecommunication networks, reaching critical mass means achieving enough users so that revenues cover costs and enough people can contribute interactively to the production and transmission of information to ensure the vitality and quality of the service - thus attracting further users. Additionally, when coverage of such services has expanded sufficiently, current non-users are likely to have nearby users whom they can emulate.

In Britain and Germany, the videotex suppliers attempted to reach critical mass in ways which differed from the direct method of terminal distribution used in France. Firstly they tried a 'management of expectations' strategy (in the popular music industry this might be put more succinctly as 'hype') by listing possible applications for the system and forecasting that within a short time an enormous number of people would in fact be using videotex for such applications. When this failed, the system suppliers in both countries turned to 'target marketing'. They reasoned that a kind of critical mass can be achieved within particular economic and social communities and sub-groups: the first groups should be those with a clear need for the unique features of videotex combined with the purchasing power to pay for it, and then the revenues from these first groups could fund the targeting of groups with similar but initially less articulated needs, thus provoking a 'chain reaction' or a 'trickle-down' effect from early to late adopters.

A certain amount of success was achieved in this way in Germany and Britain, notably in the travel sector, but as a strategy for increasing overall usage it has run a poor second to the direct 'implantation' strategy, coupled with service flexibility, seen in France. There it was reasoned that direct distribution, coupled with the provision of at least one good reason to use the system (the Electronic Directory) and the freedom for other organisations to produce new services to the design they thought best, would be the best way to stimulate demand.

The Framework of Innovation: Industry Structures and Policy Processes

The above lessons from the study of innovation practices may go some way towards explaining the differential success of the three systems at the time of their launch and shortly afterwards. But, as can be seen throughout the national studies in this book, relative success can change over time, and initial failures can be turned into success (as well as vice versa) by changes in approach and design and by adaptation to new expressions of demand and environmental factors. The videotex systems in the UK, France and Germany have all been running for some time, so it is legitimate to ask questions like: why has British Telecom not emulated the successful French model and relaunched Prestel on a different network and with a different terminal distribution programme? Why did the DBP recapitulate so many of the mistakes made in the UK even though it launched its system four years later, after Prestel had clearly been seen to confound its high expectations? Are there any clouds on the horizon of Télétel which might overshadow its success thus far?

Some answers to these questions may be attempted with reference to the environmental 'framework conditions' in which the suppliers of the three videotex systems were working. These videotex systems were clearly not simply a technical development, but also an economic and social development, which means that they were developed within specific industrial and social structures, co-ordinated and guided by governance mechanisms which include both market forces and political constraints and regulations. While the three suppliers were similar insofar as they were all publicly-owned telecommunications providers, they were in fact operating within a relatively diverse set of constraints which affected both the initial conceptions and the outcomes of their videotex systems. The most crucial areas in which the effects of these differing environments could be seen were those of integration and control.

When Prestel was being developed, the UK Post Office was already moving away from being a state agency towards a more market-oriented 'state corporation' and Prestel indeed was set up as a separate 'profit centre', prefiguring the moves of the 1979 Conservative government towards privatisation and deregulation of public monopolies. Even before this, the need to gain allies and stifle potential opposition had forced a parcelling out of the efforts and projected rewards of videotex development which led to the Post Office's loss of control of crucial assets in the innovation process. By the time the failure of the initial conception of Prestel became evident, the end of the Post Office monopolies of telecommunications services and their underlying networks was clear. Private videotex systems had appeared almost as soon as Prestel was launched, and soon went beyond their initial 'in-house' uses following 1981 legislation which liberalised value-added services. Later, when the need to relaunch Prestel became urgent, it was starved of investment money so that the pre-privatisation balance sheet of BT would look more positive. The privatised BT was then prohibited from practising the kind of cross-subsidisation between network, equipment and service provision which was an essential part of the French programme. For instance, the part of BT which runs Prestel is not allowed to count the extra local telephone traffic generated by videotex as part of the plus side of its balance sheet, even though the cost of the Prestel access network is debited from it.

By the time British Telecom seriously investigated the possibility of a Télétel-like system in 1988, the company was faced with the probability that the services which would be made possible by such a system might well also be accessed via the networks of its rivals. Thus the overall economic return to BT of what would have to be a massive investment was made even more uncertain. So, in addition to a psychological barrier against the promotion of value-added services to households - the result of the original, dramatic failure of Prestel to attract its forecast market - the regulatory environment created by the process of liberalising telecommunications in the UK has made investment in new, large-scale videotex systems an unlikely option in the near future.

In Germany, the answer to the question of why Bildschirmtext repeated many of Prestel's mistakes can also be found in the structure of the regulatory environment in which it was developed. The bitter legal disputes between the federal government and the Länder delayed the start of btx, and might have provided a 'breathing space' for learning Prestel's lessons and redesigning the system. True, there was some redesign, embodied in the semi-decentralised IBM system. But the scope of the changes was limited because the power struggle between Bund and Länder was fought in terms of how the 'essential nature' of videotex was conceived. To have changed the basic design of the system in any radical way would have meant reopening the debate and perhaps thus altering the outcome. Additionally, some of the most important supporters of btx, the representatives of the small business sector, were in favour of centralised, publicly-owned database hosts for the above-mentioned 'democratising' reasons.

Some improvements on the Prestel concept did indeed take place, for instance the incorporation of gatewayed external computers from the beginning of the service, the outward-looking link to the French system, and even the choice of the CEPT 1 display standard (initially a disadvantage because of slow screen setup time, but becoming an attraction as fast ISDN access diffuses across the country). The Bundespost was able to adjust its policy on terminal distribution by commissioning its cheap Multitels and Bitels, and a certain amount of liberalisation of modem provision contributed to the growth of PC-based terminals too. Crucially, the continuing monopoly on data communications networks enjoyed by the DBP enabled it to maintain a price differential between btx and other forms of data communication, thus making videotex an attractive option for more firms and sectors than in the UK. It remains to be seen whether btx can reach a large enough mass to maintain momentum in the face of current and forthcoming telecoms liberalisation moves.

In France, the DGT was working within a different tradition, and developed a concept of videotex that was distinctly different from that conceived in Britain and Germany. The tradition of government-inspired 'grands projets' ensured adequate resources for Télétel once the decision had been taken to construct the system. But paradoxically, the success of Télétel may have been due to the fact that the DGT's concept of videotex was in some ways less ambitious than that of the Post Offices in Britain and Germany. While Prestel and Bildschirmtext were seen from the start as creating new information services for the public, managed and designed by the system providers, Télétel was originally a more restricted concept - one perhaps more congruent with the DGT's function as a telecommunications network administration.

Télétel aimed to provide a substitute for the printed telephone directory, and to help justify the creation of a large-scale data communications network. Through the commissioning of large orders for terminals it would provide a boost for the French electronics industry - without forcing that industry to bear a marketing burden which would make a return on investment uncertain. Because the DGT wished to restrict its role to that of a telecommunications network provider, it was willing to allow external organisations to design and implement their own services as long as they used the public data network: the resulting traffic increases were then very welcome, but the original internal justification for Télétel was equally focused on the economic benefits of the electronic directory. It is to the DGT's credit that it was able to exploit the flexibility which this concept provided, and allow external services to develop in ways which were not originally planned.

As in Germany, the network monopoly enjoyed by the DGT/France télécom has allowed it to retain control over the development of Télétel. As can be seen in other chapters, its success has overcome initial distrust in other countries and has led to increasingly significant exports of both the design of the system (usually together with the equipment needed to implement this design) and the services themselves. Potential problems, as for all the countries covered in this book, arise from the increasing internationalisation of value-added services, encouraged by moves to liberalise the provision of both services and networks within Europe. In addition, France télécom has to show that Télétel can adapt to technical change in display standards, transmission speeds and more intelligent use of distributed computing features, so that the huge investment in Télétel does not later become a burden inhibiting further progress.

So, the outcomes of the videotex development processes in the UK, France and Germany can at one level be seen to be the result of design, and sometimes redesign, decisions taken by the providers of those systems. More fundamentally, however, these decisions themselves were constrained by the framework of industry structures and policy processes in which they were embedded. The result was that the capability for integration and control of the many different elements needed for the success of such large-scale innovations, and the ability to adapt flexibly to signals from the users of the new systems, were not distributed in equal measure to the three system providers. It is with these issues of control, integration and flexibility that all the videotex providers in all the countries covered in this book must struggle. Because of their size (or at least the size of the telecommunications organisations providing them) and because they were in the game from the beginning, the three - or perhaps just two - models attempted in Britain, France and Germany have defined some of the basic parameters around which other countries can innovate. They have set the stage.

CHAPTER 3

ITALY: THE SLOW TAKEOFF OF AN 'UNIDENTIFIED FLYING OBJECT'

Gianpietro Mazzoleni
Department of Sociology and Political Science
University of Salerno

Introduction

The adventure of the introduction of videotex in Italy is emblematic of the peculiar way this country copes with the challenges of technological and cultural changes in the field of mass communication. At the same time, however, it is but one chapter in the troubled story that most industrially advanced countries in Europe, and elsewhere, encounter vis-à-vis the launching of this new telematic service. The remarkable exception of France where videotex has become a commercial success that many videotex operating companies in various countries are switching to its technology and diffusion patterns, provides the contrasting background that pinpoints the drawbacks of other experiences.

Italy was a fairly passive participant in the early Esprit and similar programs of technological development. At home, the Telecommunications Plan of the early '80s was judged as a "late-comer" at a moment when the rules of the game had already been set out by other multinational actors. Even today it is not sure that the Plan has met all its prefixed goals: certainly it failed its forecast of 400,000 subscribers to videotex by 1990. The mention of the videotex medium in the "Plan" appeared to be something that the Government could not help taking into consideration, since at the time a number of European countries were also working on similar projects. The good intentions listed in the Plan were not always followed by good deeds. While France pumped political and financial resources into the pursuit of the success of Minitel, Italy showed scarce enthusiasm in this (as well as in other) new service(s). It is no wonder that for a long time Videotel (as it is called in Italy) was a sort of U.F.O. in the "inter-space" of domestic telecommunications.

The missed "takeoff" of the service, however, should not be blamed merely on the short-sightedness of Italy's policy-makers as one can advance the reasonable hypothesis that in most national contexts there existed, and somehow still exists, a significant gap between the offer (rapid technological progress) and the demand (slow cultural and social adaptation).

This could well account for, among other undeniable shortcomings, the scarce success of videotex in many countries.

1. Short history of videotex development

The introduction of videotex in Italy started in 1982, by decision of the Ministry of PTT which entrusted the State owned telephone company SIP with the running of the service. As with all Western countries that considered the opportunity to introduce the new medium, the commercial phase was preceded by a field trial that lasted about four years.

The standard adopted for the experimental phase was the British Prestel, at the time the most advanced system on the marketplace. About 1,000 users in six cities were sampled for the trial: 85% were from the business sector, 15% from households. This fact reveals the original intention of the service operator to privilege the diffusion among industrial and commercial sectors. The choice for the business market was clear in the SIP's policies. In the experiment the contents were offered by 160 information providers for a total of 75,000 pages.

Unlike other countries, however, Italy did not exhibit much "political" concern for the undertaking. In fact it failed to appoint a steering committee with the task of monitoring the experimentation. Nor was a key objective the study of the social consequences of the introduction

31

H. Bouwman and M. Christoffersen (eds.), Relaunching Videotex, 31–38.
© 1992 *Kluwer Academic Publishers.*

of this new medium. The focus of the trial was almost exclusively technical and just slightly commercial.

The new service was not enthusiasticallyreceived by the users: surveys unveiled their dissatisfaction with the quality of the technology and of the information conveyed. The high costs of connection to videotex imposed by the Ministry did not help the effort to make it popular among potential subscribers. At the same time the early Information Providers displayed some initial resistance to the call to make investments, as it did not appear clear to them the extent and quality of profit returns. The result of the trial was a total flop which significantly discouraged SIP from passing quickly to the phase of regular service. Eventually, in 1986, the Ministry of PTT issued three decrees that officially inaugurated the commercialization of the "Videotel". They confirmed SIP as the sole system operator and liberalized the marketing of the terminals (both dedicated and in general) previously monopolized by the public telephone company. They also set the fees for professional customers and residential users. The standard chosen was again the CEPT, level 3 (Prestel). Regular service started in June of the same year.

It took a couple of years of practically no newsworthiness on the part of Videotel before SIP signalled serious interest in its takeoff as a potential service for a mass market. The company realized that the success of the new service was tied to a radical revision of its technical and commercial "pillars". Accordingly, access to a multistandard system was made possible, enabling availability on the terminals of all three existing standards. Even though no official dismissal of the Prestel standard has been decided upon, SIP itself operates on a regular basis with the French standard Télétel which it considers the sole solution with a future.

On the commercial front SIP partially imitated the French practice of providing the subscribers with a dedicated set (which incorporates the modem and, optionally, the printer), renting it for a few thousand lire a month. More recently the operating telecom company urged Government authorities to abolish the "entrance" tax (the annual fee), that had revealed itself a stumbling block to a wider diffusion of Videotel. Beginning from January 1, 1991, the only costs of Videotel are those for the actual use of the services.

2. General description of the current videotex situation

The Videotel service is currently at a turning point in its development. After several years of stagnation, according to SIP authorities its diffusion is about to register a sharp increase in the next few years. The signals of better times ahead are to be found in the recent favorable figures of the number of subscribers and in the range of services offered by the information providers: the impression is that videotex is becoming more well known to a larger population than it has been in the recent past, with the result that more firms look at the service as a field to invest in with profit.

The identity card of the Italian videotex lists the following features:

CONFIGURATION OF THE SYSTEM

Apart from the provision of the electronic telephone directory, SIP acts mostly as a common carrier and a provider of switching, transmission and administration. The host system of the service is made of a number of computers, which are SIP property, located in the cities of Milan (for the North), Bologna (for the Center) and Rome (for the South and Islands). There is also a series of private external hosts/computers which are in control of some auxiliary services channelled through the SIP Videotel network (X25 dedicated) and connected gateway to the system via the public Packet Switched Network Itapac. (See Fig. 1)

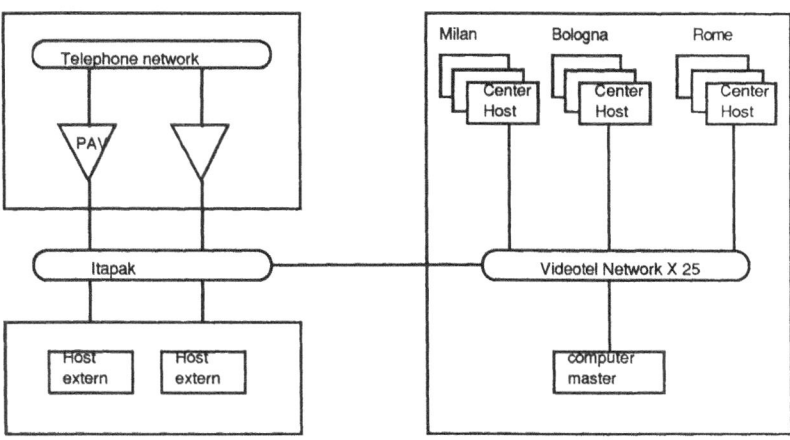

Figure 1: Italian Videotex Network

The service is of a multistandard kind as it employs all three standards developed in Europe: CEPT level 1 (Bildschmirtext), 2 (Télétel) and 3 (Prestel). The user may have access to any of them, depending on the type of terminal he/she owns. Independently of the terminal used, however, the transmission speed of the modem is 1200/75 bit/s.

COST STRUCTURE

The tariffs for access to the Videotel are defined by the Ministry of PTT in a specific government decree.
There is no longer a subscription fee. The charges are:
- 7,000 lire (= 4.6 ECU) monthly for the rental of the terminal (the tariff includes the cost for the technical maintenance)
- 12,000 lire (= 7.8 ECU) yearly, for "technical surveillance"
- about 200 lire (= 0.13 ECU) the cost of a unit at the moment of connection to the telephone line
- 150 lire (=0.10 ECU) every three minutes of connection, from 8 am to 10 pm on week days
- 150 lire every nine minutes of connection, at night and on weekends
- user fees depending on the types of services accessed and determined by the IPs, based either on per page or per time pattern. In no case should the per page service exceed 9,900 lire (= 6.5 ECU).

The charges are registered on the ordinary telephone bill and are collected by SIP on behalf of the IPs. The company withholds 10% of said income.
Beginning January 1, 1992 the "kiosque" system will be introduced. That will imply the adoption of the French Télétel solution, i.e. the time fee instead of the page fee, and the abandonment of the cumbersome password system that presently allows access to the service.

AMOUNT AND TYPE OF IP'S AND SERVICES OFFERED

As of June 30, 1991 the number of Information Providers of Italian Videotel was 933, about 300 of which connected gateway: an increase of 100% since June of 1990. This confirms the aforesaid positive trend that the service has taken in the last few months.
Videotel offers a whole range of services (about 2,300) that cover practically every area: commercial, cultural, educational, informational, administrative etc.

A brief sample follows:

- teleshopping
- tourist agencies
- consultancy to consumers
- hotel reservation
- environment
- yearbooks
- stock exchange
- football
- cinema
- classifications and survey data
- sporting activities
- auto, boat, cycles sales and rentals
- conventions

- medical consultancy
- mailbox
- finance and taxes
- gastronomy
- games and tests
- telematic newspapers
- homebanking
- laws
- expertises on line
- telesoftware
- transportation timetables
- transactions (various types)
- billing facilities

NUMBER OF USERS, TRAFFIC STATISTICS

The number of subscribers has also increased remarkably in the past twelve months, especially since SIP offered the terminals at a inexpensive rate[7] and since the Ministry of PTT abolished the standing charge in August 1990. Potential users have begun to consider the convenience of the new medium and are applying for it at an average monthly rate of 2,000. On May 31, 1991 the total number of subscribers to Videotel was 155,584. There were 147,136 subscribers on December 31, 1990: a number that covered only 52.9% of SIP's overly optimistic forecast! The company predicts to reach 2 million users by 1995[8]. The average time spent by users in front of the Videotel terminals was 35 minutes in May 1991 (26 minutes in December 1990).
The breakdown of the users in 1991 shows that 30% are residential and 70% professional.
There were 144,000 terminals rented by SIP in May 1991. The following table summarizes the basic statistics of the use of the service.

---

[7] The rental of dedicated terminals at a low fee represents an economic loss to SIP. The telecom company amortizes it in 5 years! Of course there exists a cross-subsidization that allows SIP not to register the loss on the balance sheet. It is indeed what in Italy is commonly called a "political fee" that SIP sustains in order to launch the service.

[8] The total Italian population is 57 million; the number of households approx. 20 million.

TABLE 1. Number of users of Videotel 1986-1991. Number of terminals.
Average connection time per session (in minutes). Total calls in a year

	Users	Terminals	Time (')	N. of calls
1986	5000	n.a.	n.a.	n.a.
1987	3000	n.a.	n.a.	n.a.
1988	27499	n.a.	16.0	371574
1989	80339	40004	23.0	3105280
1990	147136	133330	26.1	9868624
1991*	155584	144012	35.7	4984456

(*) Up to May 31.   Source: SIP

3. Policy analysis

In the story of Italy's videotex there are two main actors: the Government and the public telephone company SIP. A significant actor in other countries, national industry, is almost entirely absent here. Such a situation has an important impact on the general characteristics of the public policies regarding videotex (as well as of other new technologies). The "low profile", so to speak, of the attitude of governmental authorities and service operators on the issue of videotex displayed in the early years is probably due to, besides the aforesaid superstructural reasons, the lack of a driving force of an interested electronic industry in the national arena. While in some other national contexts (eg. France, Great Britain, Denmark) the presence and the pressure of the industrial sector was decisive and/or very influential in deciding the direction that the videotex adventure has taken there. In Italy the issue of the computers provided, say, by Olivetti or the terminals provided by some other leading manufacturer was never on the public agenda of policy-makers. In other words, the lack of incentive to boost domestic hardware producers has taken away the incentive to engage in a national way in videotex.

The immediate result was that of the passive, hetero-directed general policy stand, meaning that Italian developers relied extensively on what was being developed abroad, from the technical standards to the patterns of the field test to the blueprints of the information conveyed, up to the recent switch to a Télétel solution.

The Government, as institutional policy-maker, has shown intensive productivity (from a quantitative perspective) in issuing administrative decrees to inaugurate videotex. Beginning from the first Decree of October 30, 1982 to the last decree of August 7, 1990 they represent approx. 10 pronouncements specifically dedicated to the new medium, besides a number of those indirectly related to it, regarding, for example, the tariffs of the network Itapac, and other telecom areas.

From a qualitative (or content) standpoint this bureaucratic production fails to show the clear motives and policy goals of the Government. As has been mentioned, videotex was listed in the early Telecommunications Plan among other services, with no special emphasis. Thus it is no wonder that the official policy documents (i.e. the Government's decrees) far from signal any concern of the State decision makers regarding videotel. In the first half of the past decade, the philosophy of public administration was still the bare defense of the traditional State monopoly of both hardware and software in the sectors of telecom (and of mass media). Accordingly, private operators were banned from running videotex services, the public telephone company was the only licensee, the terminals could only be provided by SIP, and other minor restrictions revealed the prevailing monopolistic outlook. Starting in 1984, a slow and gradual loosening of the centralized control on all matters regarding videotex can be detected. Even if the Ministry of PTT did not renounce detailing norms and restrictions in the ensuing decrees, it implemented what one may call a "soft deregulation", consistent with the change in the general political mood toward the role of the State vis-à-vis the communications media. That was quite clear in the broadcasting sector, where the State's monopolistic stance had been de facto defeated by the establishment of a strong commercial, private television system. With regard to videotex, the "deregulation" concerned the marketing of the terminals, the possibility for private companies to establish videotex services, and the abolition of the annual tax. Surely this cannot be regarded as a dramatic liberalization but it is objectively the maximum that a centralized administration of the European kind, jealous of its control over the telephone medium, can concede.

SIP, the other main actor in the policy-arena, as sole licensee of the State's telephonic service, hardly demurs the monopoly-oriented attitude of the central administration. However, having also a commercial nature it is in a better position to sense the changing patterns of the marketplace. So, it did not obstruct the Ministry when it liberalized the supply of terminals in 1984, and it purposely pressed the Government to abolish the videotex tax. For these and other reasons that may be too complicated to illustrate here, one can observe that to a certain extent the public telephone company played the dominant role in the videotex story. The subaltern conduct of the Government on this issue can be explained in terms of the ancillary role that it played with respect to SIP. In other words, the telephone company took on the task of setting out the policies necessary to inaugurate the new service. The Government's policy action was thus the official translation into administrative forms of the blueprints worked out by SIP policy-makers. SIP, as has been noted, did not show itself to believe excessively in the success of videotex in Italy from the very beginning. The same attitude is mirrored in the Government's decrees that say nothing to create the impression that central policy-makers are investing political and economic resources to assure the takeoff of this service.

The passive position of public administration has left a large margin of autonomy to SIP in implementing their policies regarding videotex. The recent adoption of the standard CEPT level 2 (Télétel), in contrast with the 1986 choice of the Government in favor of the sole Prestel, is the most clear example of such autonomy. It may well be that soon the Ministry of PTT will issue a decree that allows the "multistandard" solution. The fact remains that SIP anticipated the official policy, and in doing so, demonstrated itself to be the dominant actor in the videotex undertaking.

With regard to the practical shift to the French standard, (dating back to 1988) this was caused by the realization on the part of the domestic videotex operators that the existing policies and technical options of the Prestel "scenario" were largely unfit to guarantee a notable commercial fortune of the service. The success achieved by the French model eventually won from the skepticism of Italian operators who started to believe in a future for videotex, probably for the first time. In fact, the turning point was set in 1988-1989, as described above. From that moment the commercial policies of SIP also began to change and to become more aggressive, with some evident positive returns. The reorientation in SIP's policy approach toward videotex is based exclusively on market-wise considerations after a long period of policy elaboration and implementation quite insensitive to the actual response of the final consumer of the new service.

As a public operator, SIP interacts with other key actors, such as the Information Providers. Especially in the first, founding phase, the IPs collaborated creatively in launching Videotel. In a way one can say that they were the only optimists, enthusiastically accepting to invest in the adventure. ANFoV (National Association of Information Providers) acted as a catalyst between the telephone company and the market of information potential suppliers. It stimulated a larger involvement of its associates and contacted further investors. It also took part in the policymaking process working out an ethical code regarding such key issues as copyright, security, and privacy the last of which was taken into consideration by the Government and political parties for legislation to be discussed in Parliament. In the second phase, at the beginning of the regular service and before the shift to the French standard, the collaboration between ANFoV and SIP reached very low levels. The IPs realized that the telecom company were indifferent toward the new service. SIP was not pulling, Videotel was not expanding, consequently several IPs, not seeing profits on the horizon, stopped investing and many others quit altogether. The poor content channelled on videotex was all that was needed to secure its decent diffusion. That eventually impoverished the image and quality of Videotel.

To conclude these notes on the policy front, a few more comments on the absence of the electronic industry in the arena. There exists a paradox in the whole story of "mass telematics" Italian-style. As early as 1982, coinciding with the experimental inauguration of both videotex and teletext, the Government passed Law n. 63 which intended to restructure and rationalize the country's ailing consumer electronics industry. It was a sector that produced mostly television sets (and few other consumer goods). There were no plants producing videorecorders, terminals, decoders, modems. That law however a solemn policy act by public planners, hardly refers to or takes heed of the opportunities that the launching of the two new telematic media could provide toward revitalizing an industrial sector.

One emblematic example of the contradiction in public policies regarding the new media was the two-year delay in choosing a standard for teletext and in the meantime the prohibition of marketing receivers with decoders, thus forcing the domestic electronic industry to a standstill that eventually favored foreign manufacturers. A weak industrial sector such as the Italian one has been for much of the past decade could hardly afford to exert influence on the process of policy-making in the field of mass telematics.

On the whole one may conclude that Italy's principal actors in the videotex policy arena displayed a perfect bureaucratic approach to the whole matter. In fact no technical and normative aspects were left unnoticed. What was missing was the confidence in the goodness of an adventure that seemed alien to the expectations of a class of policy-makers definitely more interested in the defense of the status quo than in finding solutions for consumers demands and problems.

4. Conclusion and future perspectives

The analysis of the problems that have risen with the introduction of videotex in Italy gives an account of the scarce diffusion that the new medium has registered so far in the domestic marketplace. The responsibility of the main policy-makers is undeniable: their lack of belief in the medium was a major hindrance to the promotion of it among potential consumers. Regarding the users, undoubtedly at the beginning of the videotex adventure there existed little if any interest in the new service. Either for general economic reasons (the recession of the '70s had brought austerity in family spending in Italy), or for cultural reasons (the market was at the time 'distracted' by booming commercial broadcasting), or for a series of various reasons (the limited diffusion of informatic literacy), the extent of the demand for new media technologies (among other things quite unknown to most) was rather poor and frail. Existing under such conditions, Videotel could hardly get off the ground.

Looking ahead instead, there are signs that a reasonable conversion both in the policy-makers' outlook and in the consumers' readiness to accept the product is taking place. Again the French example is paradigma tic, as it confutes the pessimism of the domestic videotex operators and encourages investors and consumers to "try again". At the turn of the decade, the Ministry of PTT and SIP were suddenly "enlightened" and started to remove the obsolete chains that choked the growth of the new service. To reach the 1995 target of 2 million users, SIP has made a substantial financial investment of 1,000 billion lire (1 ECU = 1,536 lire), which dramatically marks the shift in the policy perspective of the public videotex operator: from lukewarm faith to wholehearted belief!

It may be impossible to reach 2 million subscriptions from today's 156,000 in just four years. An apparently unreachable target however demonstrates the determination of the company to place all available forces and resources into the enterprise. SIP has decided to take on the risky challenge, and in order to win it, is about to introduce further, major innovations in Videotel hard and software. The service as it is offered today is still too "user unfriendly" even to customers familiar with computers. A condition for its mass diffusion is thus the elimination of all the stumbling blocks that impede a smooth dialogue between user and conveyed information. As noted, the customers have started to show reassuring feedback regarding the first changes in the technical design and commercial packaging of Videotel. The last, but not the least condition for success is the further reduction of the cost of the service. Even though economic austerity is out of date in Italy's collective imagery, an expensive videotex could well lose in the competition with other "new media" (teletext, homevideo, DBS satellite, other high-tech domestic hardware) on the cost factor alone. Italians are already massively showing their preference for the latter commodities: these are produced as perfect "consumer goods" as they rarely challenge the users' technical competence.

Videotel has reached a critical point: either it turns into a real "consumer good" or it is destined to drop out of the market completely. Either it reveals its friendly and practical identity or it will remain the crippled U.F.O. that, following a toilsome landing, is unable to take off again.

We now have a pilot with good intentions. Let us see if he can get the spaceship off the ground by 1995!

References

SIP, **Il sistema videotel**, Rome.

VT, **Nuova Comunicazione-Videotel International Review**, I, (6 June 1991).

Grazzini, E. 'La SIP dopo il fallimento tenta il rilancio. Il Videotel? E'un videoflop', **Il Corriere della Sera** (20 September 1989).

Grazzini, E. 'Videotel, affari sulla rete', **Il Corriere della Sera** (30 June 1990).

'Verso il boom del Videotel SIP. Nasce il dialogo in tempo reale', **Il Corriere della Sera** (5 July 1990).

Mazzoleni, G. (1986)'Italy'. In H. J. Kleinsteuber, D. McQuail, K. Siune (eds.), **Electronic Media and Politics in Western Europe**. Frankfurt, Campus, pp. 169-190.

Mazzoleni, G. (1986) 'Mass Telematics. Facts and Fiction'. In D. McQuail, K. Siune (eds.), **New Media Politics**. London, Sage, pp. 100-114.

CHAPTER 4

THE NETHERLANDS: BUNDLING SUCCESSES OR BUNDLING FAILURES? THE ART OF SYSTEM INTEGRATION

Harry Bouwman
Department of Communication,
University of Amsterdam

Wim Hulsink
Rotterdam School of Management
Erasmus University Rotterdam

This case study on the introduction of videotex in the Netherlands demonstrates that the introduction of tele-information services is a complex proces during which certain collective action problems have to be solved. A successful diffusion of videotex requires besides innovations in infrastructure, equipment, services development and socio-cultural acceptance, an (inter)organizational innovation which can be achieved by integrating the separate functions of information provision, hard- and software supply and network operation. Of the first videotex systems developed in the early eighties in two cases all three roles were fulfilled by one organization: the Dutch PTT with its Viditel-system and the publisher VNU with its Ditzitel-system. Neither however have been very successful.

Recently a fourth function has been acknowledged as playing a crucial role in the realization of videotex, namely **system integration**. The system operator markets the videotex system and determines the conditions under which service providers may employ the system and consumers may have access to it. However, cooperation with regards to the diffusion of videotex has been a painful process, in which the politics of uncertainty has prevailed (Quelch & Yip, 1985). First, cooperation in the development of videotex is rather complicated and requires a fine-tuning of activities. The critical functions of system operator, network provider and information supplier have to be performed simultaneously. Secondly, videotex is not a new technology, but a new organizational combination of existing technologies, which means that the aforementioned participants are also potential competitors with conflicting interests. Thirdly in a conflictuous situation surrounding a promising market one often finds participants that only want to be part of the game in order not to miss out on an important development. Such participants may have a retarding or even a sabotaging influence if their only goal is to prevent the undertaking from becoming a success.

The Dutch case is an illustration of the laborious process involved in the evolution of (system) cooperation in the videotex industry; it took about ten years to create a consortium, in which all the organizations concerned and their different functions were integrated. After several attempts in experimental form on the part of PTT, publishers and others to provide the videotex system on their own with a minimum of mutual adjustment, a kind of integration of the various systems and cooperation among the parties involved was achieved in 1989. There has been a development towards system cooperation, in which the various tasks - information and service provision, network operation, hardware and software suppletion and system operation - have been properly divided among the various specialized actors and firms and in which coordination between the internal and external parties has been realized by the network operator of Videotex Nederland. But whether the collaborative accomplishment preludes a successful penetration in the consumer market, or whether it is merely a 'bundling of failures' in a market which is still in the proces of taking on its definite shape, has yet to be seen.

H. Bouwman and M. Christoffersen (eds.), Relaunching Videotex, 39–51.

1   The current videotex situation

Following field experiments videotex was introduced in the Netherlands in 1981. The system was called Viditel and was based on the Prestel concept. Viditel used the telephone network for sender and return channel. Initially the PTT used only one Viditel switching centre in Rotterdam. One major disadvantage was that access for both information suppliers and users was rather expensive. Users outside Rotterdam were charged long distance rates and information providers were bound to the PTT's centralized and costly approach. Recently the PTT has decided to adopt a more decentralized infrastructure. The change in configuration reflects a change in the marketing vision regarding Viditel. At the moment the PTT regards Viditel mainly as a generic facility. However Viditel has not succeeded in attracting a sufficient amount of subscribers. In 1981 Viditel began with 4000 subscribers and 130 information suppliers. In 1987 the maximun of 27.000 subscribers and 271 information providers was reached, after which the figures declined.

With the relaunching of videotex in 1990 by Videotex Nederland information providers have changed to the new Télétel-like system. It is no longer possible to subscribe to videotex. The billing system is transparent and based on a kiosk-like system. Four numbers give access to videotex services (06-7100 local telephone charge, 06-7300- 23.5 cents/minute, 37.5 cents/minute and 06-7500 50 cents/minute). The costs are included in the telephone bill, but only specified on request. The infrastructure is quite complex (see figure 1). There are three different ways to get access to Videotex Nederland the first by two-way cable (introduced in South-Limburg), the second by a hybrid variant: using a push-button telephone in combination with the teletext facility of a television-set (Rits) and the third by a Minitel-like vtx-terminal (695 guilders/300 Ecu) or Pc + modem (345 guilders/150 Ecu) + software (25 guilders/11 Ecu). Until now 30.000 Minitel-like terminals have been sold.

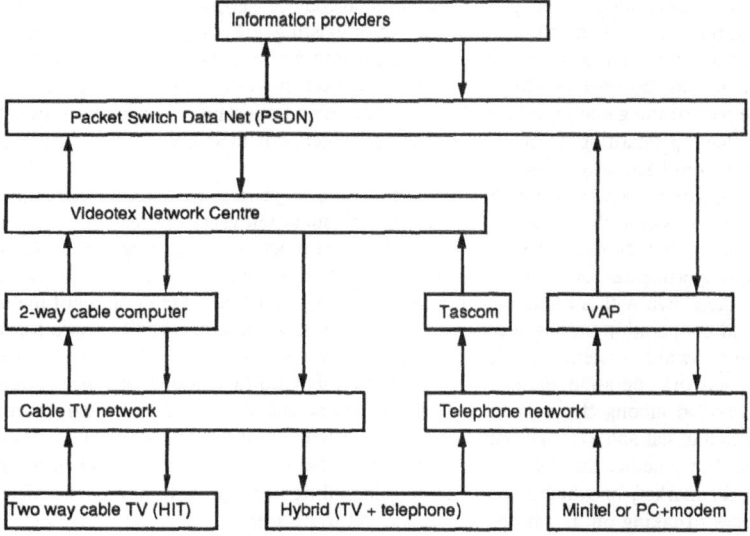

Figure 1: Configuration VTX-NL

Videotex Nederland works with international standards such as Antiope, ASCII, X 29 and Prestel 2.2+. This makes international connections possible.

Videotex Nederland offers 147 different home-banking services, transaction, information and communication services for the consumer and business market. Several services are offered at different tariffs. Information suppliers pay 750 guilders (325 Ecu) annually for the registration of the service. Of Videotex Nederland's two lowest tariffs (06-7100 and 06-7300) no amount is paid to the information providers. When the 06-7400 tariff is used the information provider receives 12 cts/min and in the case of 06-7500, 23,5 cts/min.

Table 1. Overview of services offered

Tariff	Information Services	Communication Services	Transaction Services	Home Banking
06-7100	5	-	1	-
06-7300	12	-	1	1
06-7400	106	15	19	-
06-7500	15	-	3	-

Marketing efforts have mainly been directed toward (segments of) the business market. The expectation is that 3/4 of the traffic will be generated from the business market. This is 300 million minutes in 1993. Consumers use of the videotex system is expected to rise to 100 million minutes. Table 2 illustrates the development in the amount of calls (with an average connect-time of 6 minutes). In june 1991 the average usage per user per month was 22.5 minutes (France 60 minutes). The most calls are based on the Minitel or PC-modem variant. About 8.000 users make use of the hybrid Rits approach.

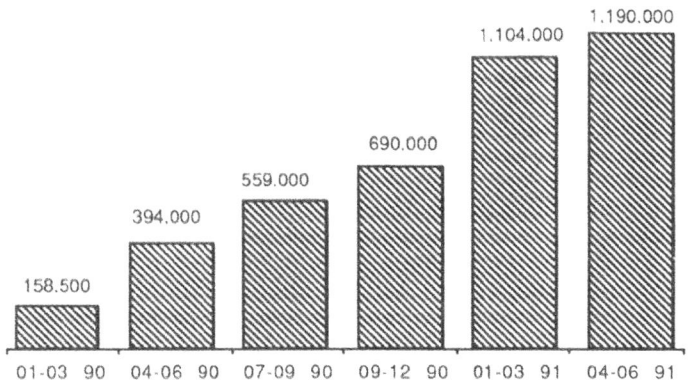

Figure 2: Number of calls per quarter 1990-june 1991

There is a steady growth of connecttime. However it is not clear if the critical mass of users, estimated at 180.000 users (Bouwman & Slaa, 1991) and services has yet been reached.

In the next paragraphs we will establish the main actors involved in the introduction of videotex, their goals, the means they used to reach their goals and which of them dominated the policy arena during various phases. We will first deal with the introduction of Viditel and secondly with the emergence of different initiatives which led to Videotex Nederland.

2. Introduction of Videotex in the Netherlands: Viditel

The Dutch PTT made a quick start with their videotex-experiment, called Viditel. In April 1978, the Cabinet sanctioned the request of the PTT to start an experiment with interactive videotex. The Cabinet stipulated that the experiment should be accompanied by a 'Steering Committee Viditel', in which the various (potential) stake-holders were represented. Its goal was to gather information and give advice regarding the judicial, technical, ergonomic and socio-economic aspects of videotex (Stuurgroep Viditel 1980). The following institutions were represented in the Steering Committtee, which was installed in september 1978: the PTT; various Ministries, including the Ministry of Interior; the Ministry of Justice; the Ministry of Education; the Ministry of Welfare, Health and Cultural Affairs; the Ministry of Science Policy and of Transport and Public Works; the National Broadcasting Organization 'NOS' and the Modern Media Foundation SMM (a branch association representing the publishers). It is of interest to note the absence of the hard- and software suppliers (e.g. Philips). The Steering Committe, chaired by a PTT-executive, was dominated by governmental agencies. The Modern Media Foundation (SMM) was the only representative of private interests amidst representatives of the public interest. The rather optimistic expectations of a convergence between public and private interests and systems cooperation in videotex were in fact, at the beginning of videotex's development in the Netherlands, an understatement for the dominance of public roles, in particular the PTT-executives.

Unaware which information providers should be involved in the experiment, the PTT originally considered the traditional print publishers, organized in the Modern Media Foundation, as the representatives of the 'electronic' information providers. There existed within the Modern Media Foundation internal problems regarding the participation in the Steering Committee between the representational organization - SMM - and the represented branch organisations: the bookpublishers, the publishers of magazines and the press. These branch associations refused an indirect participation in the Viditel-experiment via the Modern Media Foundation. They wanted to have direct access.

The SMM began as representative of the publishers in order to negotiate a contract with the PTT regarding the provision of information, the code of conduct, and the coordination of system marketing for Viditel. During these negotiations problems emerged between the publishers and the PTT because the PTT had taken all matters into its own hands: the PTT had become network provider, operator, information supplier and regulator all in one. The publishers rejected this concentration of power by the PTT and argued against it before for a Commission of Arbitration which was established in order to deal with controversial issues such as this one. This proposal was refused by the PTT. During these negotiations it became clear that the print publishers could not be considered the representatives of all the information providers. Besides the traditional information providers, there emerged a new group of information providers, for example the Yellow Pages, travel agencies, banks, mail order firms etc, that recognized videotex as an ideal vehicle for developing new services such as tele-shopping, tele-banking, tele-reservation etc. The traditional publishers looked upon videotex with mixed feelings: on the one hand they felt it was a threat to their information supplying monopoly and on the other hand they perceived videotex as an instrument for enlarging the readership of newspapers and consequently as a potentially new source of profit.

Gradually, the need was felt both in the 'traditional' publishing community and in the 'relatively new' information technology community, to combine their efforts into one collective lobby organization. In 1980 the National Information Providers Association VNVI, after 1986 known as NVI, was founded. The PTT recognized the VNVI as the representative body for the information providers of Viditel. Officially the VNVI took over the contract negotiations from the Modern Media Foundation. The discussion regarding representational problems resulted in the installation of a new Steering Committee with more balance between public and private roles. The Modern Media Foundation was replaced in the Steering Committee by representatives of the three branche organizations of print publishers (the bookpublishers, the publishers of magazines and the press) and representatives of the NVI. The public and private interests within the Steering Group became even more balanced when the National Consumer Organization 'Consumentenbond' and the Union of Journalists were also represented. The Ministries of Justice and of Education and Sciences left the Steering Committee due to lack of interest. The Steering Committee was named after its chairman Zoutendijk.

Between the first and the second interim report (1980-1981) the composition and the mandate of the Zoutendijk Committee was modified. Its aim broadened in the second interim report. Besides monitoring, supervising and evaluating the experiment, it was expected to give a recommendation for regulating Viditel (Stuurgroep Viditel 1981). The question to be answered was whether videotex was to be regulated within the legal framework of broadcasting or telecommunications or whether special legislation (a kind of 'lex specialis') was required. After reorganizing the Committee and broadening its mandate the official aim of the Zoutendijk Committee became:
- guiding the Viditel experiment, particularly in terms of its socio-economic aspects;
- giving advice concerning any rules for the application of videotex in the Netherlands;
- giving advice concerning the final introduction of Viditel in the Netherlands, and making recommendations concerning the administrative adaptation (Stuurgroep Viditel 1981).

The emphasis changed implicitly from the question - whether Viditel had to be introduced or not - to - how Viditel was to be introduced.

The publishers supported by the NVI refused placement of videotex under the Broadcasting Act, because this would exclude the participation of publishers. They preferred either self-regulation or placement under the Telecommunications Act, which would guarantee them free access to the new media. Another point of criticism, mentioned by the NVI, was the conflict of interest and the (potential) abuse of power by the PTT by way of financial, technical and commercial cross-subsidization. The VNVI argued that the PTT subsidized her activities with Viditel by means of public resources from the monopoly sector. Hence the VNVI requested for a judicial separation of the public utility and the commercial function of the PTT (VNVI 1981). The Consumer Organization had an eye for the quality and affordability of the service and the priorities of the Union of Journalists were 1) editorial independence, 2) rules for free access and 3) the quality of the information offered.

The publishers, the information providers, the consumer organization and the journalists developed a strong and sophisticated opinion regarding the role the PTT should play in Viditel. For them **any** PTT interference with the information content had to be excluded and the different PTT functions, such as service operator and common carrier, needed to be clearly defined. Pricing procedures had to be established. At the beginning of the experiment, the PTT wanted the function of network supplier only, but toward the end of the experiment it had become the operator and one of the main information providers of Viditel. In the end PTT had control over the telephone network and the Viditel-computer, PTT led the contract negotiations with the information providers and did the administration of the Viditel service. Within the Viditel-project the PTT was player, manager/coach and referee all at the same time.

The hard- and software suppliers had withdrawn from the policy making process. This was partly due to the fact that they were not represented in the Zoutendijk Committee, which meant that they did not have any direct access to the policy making process. But during the experiment they also played a waiting game. The PTT's imposing of very restrictive regulations on the necesary equipment like telephones, printers and terminals,led to the situation that the hardware suppliers had difficulties delivering high quality terminals at a reasonable price and at short notice (VNVI 1981).

Following the publication of two interim reports (June 1980 and May 1981) the Final Report of the Zoutendijk Committee was published in June 1982. The pilot experiment yielded extensive information about the technological, legal, commercial and sociological aspects of tele-information services, but the main question whether to continue or to stop the Viditel experiment was left unanswered. The Zoutendijk Committee did manage to temper the already high expectations of the PTT. The PTT was expecting more than 100,000 subscribers in 1985 while the Zoutendijk Committee did not come further than 75,000 subscribers in 1985 at the very most. According to the committee the results were due to the high costs of information and the lack of quality of the information provided. But the main problem was the choice of target groups namely the general public instead of business and professional users. The experiment made clear that Viditel was more suitable for the business market, where a well defined need for structured information could be detected. The Zoutendijk Committee urged for a definitive implementation of the Viditel-service:

"Consequently, the Steering Committee would not consider it wise to abolish Viditel on the basis of the somewhat disappointing results of the experiment. Assuming that other countries will continue in this area, the Committee fears that the Netherlands may find itself lagging behind" (Stuurgroep Viditel 1982: 14/15).

The Committee footnoted this positive recommendation with the following remarks. To increase the acceptability of the Viditel-system the Committee suggested that the costs of the information provided should be reduced while the quality, extensiveness and access to the information should be improved. The government endorsed in broad lines the recommendations of the Committee. The government agreed to the definite introduction of interactive videotex. It did not take into account the rather disappointing results of the Viditel experiment. Government policy was not based on factual experiences, but on the articulations concerning the future interest of new services and the menacing arrears regarding vidotex in the international context. The government pointed out, that the PTT was in aposition to play a special role in stimulating telecommunications. By supporting the development of new services and new markets, the government expected that the PTT would strengthen its position in the telecommunications industry.

The government considered the Viditel experiment part of an **innovation policy** in which it offered the business sector means to develop new opportunities for the hard- and software industry. Thanks to the Committee the government realized the importance of a jointventure between government and business to invest money, expertise and energy in the research, development and marketing of information technology (Tweede Kamer 1983-1984). With respect to the marketing and promotion of Viditel, the Cabinet followed the Committee's advice by adopting an active policy, directed towards the recruitment of information-intensive and economically viable industries such as transport, travel, insurance and banking.

But at the same time the government argued that the PTT should act in conformance to the market and should avoid cross-subsidization. The government deemed it necessary that a clear separation in bookkeeping between the traditional telecommmunications services (telephone and telegraph) and the new services such as Viditel be established. The argument on the other hand that experience with Viditel was indispensable for the PTT for further development of a public videotex-service, made clear that the Cabinet did not consider an organizational separation between the exploitation of Viditel and the rest of the PTT desirable.

The government wanted to stimulate Viditel, not by creating the necessary conditions for innovation through competition, but by enlarging the responsibility of the PTT that was expected to pave the way for the development of new services (Tweede Kamer 1983-1984). The government suggested that an active participation by the public administration in the Viditel-system and other similar public information systems could on the one hand stimulate the further development of tele-information systems and on the other hand embody objectives such as the administrative openness and the improvement of the efficiency of the information supply within the public administration, as well as from government to the general public. The government requested an immediate and clear-cut decision concerning the implementation of Viditel. Viditel should either be continued or liquidated. Shutting down the Viditel service meant a loss of 24 million Dfl. and would damage the development of a highly advanced telecommunications net-work; continuing with Viditel would yield the PTT indispensable experience for the further development of a public videotex infrastructure. The government expected that the Viditel service would be profitable by 1990. There was hardly any opposition against the standpoint formulated by the government. The parliamentary discussion regarding the issue which took place in 1984 lasted little over half an hour (Slaa 1987: 128) and was easily accepted by Parliament.

One could say that the decision to extend the Viditel-experiment to a permanent service was based on arguments of industrial policy. The imposed introduction of Viditel had more to do with investigating the market for tele-information services than with the high aims of technology asses-sment and socio-economic innovation. The PTT hoped to generate more traffic through its network by extending its range of telecommunications facilities; the manifest or latent needs of the general public had a lower priority. The main motive behind the PTT's offering of new services such as Viditel was the activity abroad, which for strategic reasons had to be kept pace with (Algemene Rekenkamer 1989; Slaa 1987: 129). Moreover, the PTT was interested in stimulating telephone through the exploitation of Viditel thereby optimizing the capacity of its telephone network.

Partly dictated by the disappointing results of the experiment, the PTT decided in 1984 to change the marketing focus to the professional and business markets. This change in marketing strategy resulted in a moderate, but steady annual growth in the number of (professional) subscribers and in the (professional) services provided. The PTT moved into the profitable market segments of specialized information services for finance, transport, real estate, travel agencies etc. The specialization and new strategy resulted in a gradual growth of the Viditel service, yet the number of subscribers did not meet the initial expectations. Viditel had a steady annual growth from 11,000 subscribers in 1984 to 27,000 subscribers in 1987 followed by a slight decline to 23,000 subscribers. The take off was slower than expected and consquently the break-even point for videotex was not reached.

In 1988 the National Audit Office investigated the introduction of telecommunications services by the PTT, between 1980-1987. The following services were examined: Viditel (the public videotex service), Memocom (the public electronic mail service), Datanet-1 (the public datacommunications network), the car-phone network and the green numbers service (06-phone numbers). The Government Audit Office concluded that the PTT had lost a total of approximately 287 million Dfl. on four of the five new telecommunications services. Only the green numbers service was found to be profitable, the others showed a negative result (Algemene Rekenkamer 1989). In the case of Viditel it had taken too long before technical-commercial procedures among the actors involved, necessary for a succesful exploitation, were realized. Besides the initial losses there were also unexpected costs, which had mainly to do with technological fixation, poor marketing, and bad management.

3   Relaunching Videotex: The Evolution of Cooperation

As a reaction to the near monopoly of the PTT in developing tele-information services, an uncoordinated series of competing projects was initiated: experiments with Ditzitel and with Totaalnet Zuid Limburg and plans from Telematica Infostructuren and Infodam. The aim of these often local or regional experiments was to make local government, business, education and priva-te households familiar with the new information and communication technologies. But just like in the Viditel-case, the commercial results of these initiatives lagged far behind the expectations and some of them were never realized. In 1984 VNU (one of the biggest publishers in the Netherlands) announced its own public videotex system, called **Ditzitel**. Marketed as an alternative for the PTT's Viditel, Ditzitel was designed as a low access, technically simple, hybrid system located in the city of Amsterdam. VNU initiated its Ditzitel-project on its own after the PTT, holding a monopoly on the telecommunication infrastructure, had refused to cooperate. PTT had rejected VNU's proposal for integrating the Viditel and the Ditzitel-system (Slaa 1987: 127). During the period 1984-1987 the Ditzitel concept changed several times, due to negative experiences with pilot audiences. In 1987 the entire project was cancelled before it had even gone underway. The reasons for Ditzitel's failure had to do with the PTT's refusal to adapt its infrastructure and with the problems defining the necessary software and the appropriate format which would best fit the consumers' demand. The Ditzitel-project was cancelled, according to the VNU, because 'the product developments were - in the experimental phase - considerably high, the technology was too uncertain, the time to recover the costs was considered too long'. In short the risk was too high. VNU did learn from the Ditzitel-experiment that the new electronic market did not have a high growth potential and that is was not a real threat to its market of magazines.

The **Cable experiment Zuid-Limburg** (also known as Totaalnet), an experimental videotex system located in the south of the Netherlands, was launched in 1981 by the Dutch government as the spearhead of its informatisation policy. The system was originally meant to provide a sophisticated play-ground for new consumer-oriented telecommunications services and anticipated a futuristic, interactive 'wired city' concept. Originally the planned scale of the experiment was large (in 1983 the biggest in Europe). Its most recent form was but a moderate version of the original plans; the cable networks used were only partly interactive; its potential subscriber population was approx. 100,000 households. It lasted until 1989 before the communication system actually became operational. For the first time in the Netherlands a special company was formed to realize the cable-experiment (Kabelexperiment Zuid-Limburg BV). Information providers, hard- and software industries and the Dutch PTT participated in this company. It marketed the system and served as host organization for information providers. This experiment too however, turned out to be a failure. In 1991 the Totaalnet Zuid Limburg was terminated; the actual population of subscribers/users remained far below expectations (in 1990: 1700 subscribers). The experiment showed that the development of the information society, and its service economy was technologically possible, but that the market for new eletronic services had little growth potential.

In 1986, as a follow-up to a (V)NVI initiative called Infobox, a project team was installed to study the feasibility of a large-scale introduction of videotex in the Netherlands. The project team, supported by the PTT, NVI and the soft- and hardware industries concluded in 1987 that the exploitation of a large-scale videotex service was feasible. This meant that within the not too distant future 700,000 videotex terminals (600,000 consumers and 100,000 business users) would be connected to a hybrid videotex system using cable and telephone called **Infodam**. Another experiment with videotex called **Telematica Infostructuren**, was a regional experiment for the realization of a videotex system for consumers and self-employed (e.g. farmers), based on the French Minitel/Téletel system. These two initiatives included plans for large-scale or regional introduction of consumer videotex; neither was ever fully implemented.

The cooperation between interdependent actors with convergent and divergent interests was bypassed in a series of uncoordinated and even competitive moves during various experiments with videotex. The recently adopted decentralized approach (by regionalized Videotex Access Points) is characterized by the fact that the PTT has limited its role to 'system integrator' giving more room to service providers to follow their own marketing strategy. It took a relatively long time (almost ten years) before the different actors, such as the PTT, private network suppliers, cable tv networks, information suppliers, (business) users of information, capital investors, and the like, became aware of their prisoner's dilemma and realized that more cooperation and a better coordination between the projects were necessary. Slaa (1989, 12) characterized the Dutch experiments as a mixture of coordination and cooperation on the one hand and free riding and sabotage on the other: 'a number of participants became involved in the development of a new service, not so much to actually develop this service, as much as to keep an eye on the competitors or to slow down the development'.

At long last under the guidance of the PTT, the Department of Trade and Industry and potential investors cooperation between the competing experiments emerged. On the basis of its dedicated role of network provider the PTT could play a mediating role between the various experiments. But the PTT was not only a 'neutral' referee, it was also a major player. As well as being the neutral network provider for various videotex systems, PTT still had its own Viditel-system as well. So the PTT had a relatively strong bargaining position. Having a working videotex system and given the conflicting interests between system operator (Viditel) and network provider (for the other videotex systems), the PTT could favor its own Viditel-system above the others. At the same time however the PTT realized based on ten years of experience with videotex, that cooperation with other participants and harmonization of the other videotex systems was essential. Another factor, which must have influenced the decision of the PTT to initiate joint ventures with other competitive initiatives was the slow decline in the growth of Viditel from 1988 onwards. The last factor was the privatisation of the PTT and the liberalisation of the telecommunications market in the Netherlands since the 1st of January 1990. Deregulation made it legally possible for the PTT to enter into joint ventures, but at the same time it forced the PTT to change its strategy from one of domination based on a natural monopoly to one of a more mediating and coordinating nature. Also the Ministry of Trade and Industry, that had invested a lot of money in the various experiments, demanded cooperation and harmonisation. According to the Director General of the Ministry 'there is no room for more than one large-scale system, so cooperation is necessary' (Staatscourant April 8, 1988). Other (potential) investors requested further cooperation and harmonisation of the systems in order to secure their investments.

Pressure from potential investors resulted in a demand from the Ministry of Trade and Industry and particularly the PTT-administration for a form of mutual coordination and system integration between the concepts of Viditel PTT, Telematica Infostructuren, Totaalnet Zuid Limburg, Infodam and Demos (another new videotex-system promoted by the cable operator Deltakabel). The intention to collaborate was agreed upon in September, 1989 and became operational in March, 1990. **Videotex Nederland b.v.** was established. But collaboration went even further. In 1991, two existing videotex systems were taken over: the incorporation of RITS-system (a hybrid system using the telephone lines and local television cable networks) and the Comnet videotex system (offering information and communication services) in the infrastructure of **Videotex Nederland** meant an increase of regular users (i.e. terminals in use) to approximately 100,000.

Videotex Nederland integrated all the various initiatives into one transparent network environment (see figure one); all three possible infrastructural configurations (telephone/data network, cable network an a combination of both) were applied. At the same time it was strived for that the system became fully transparent for the information provider; this meant that an information provider could deliver its information or service in one general format, while the system operator, Videotex Nederland, would take care of the necessary protocol conversions. The main shareholder in Videotex Nederland BV is PTT Telecom with 19.5 percent share; the other important shareholders are two Dutch banks, a computer firm, an institutional investor, the representative of cable operators and Intelmatique (the French marketing organisation of Télétel). Intelmatique has a 17 percent share (6.6 million guilder or 2.8 million Ecu).

System cooperation emerged because the parties involved realized their interdependence and the market and technology of videotex had matured. As such the Dutch videotex case, can be seen as an illustration of the 'evolution of cooperation' (Axelrod 1984; Hulsink 1989). Quelch and Yip (1985: 291) consider the emergence of cooperation in the videotex industry to be the result of three factors:

1) a dominant participant: the Dutch PTT;
2) the interorganizational relationships becoming more systematized and stabilized over time: - the Dutch case took more than ten years of interorganizational learning;
3) an increasing technical standardisation providing integration by installing interfaces.

The system of Videotex Nederland BV is 'a typically Dutch solution' of a combination of three different configurations based upon PTT networks (like Viditel), (interactive) cable tv networks and a hybrid variant that uses cable networks to display information and uses the telephone networks to order or to communicate information.

4   Discussion and Future Perpsectives: It's Now or Never

Telecommunications policy is the result of present technological possibilities and past agreements and designed policy solutions. The new telecommunications services emerged somewhere between the press, broadcasting and telecommunications and did not generally fit into the existing legal-institutional structures. New services were associated with existing services provided through the same infrastructure, but existing legislation was network-based and not service based. The policy question became: how can the etablished legislation be adapted in order to encompass the new technological possibilities. The introduction of videotex in the Netherlands is an illustration of this. The Dutch government was presented with a (technological) 'fait accompli', which was neglected as long as possible by focusing mainly on the vested interests of the telecommunications establishment.

As for decision making regarding new media, special attention needs to be given to the following two aspects: a) the permanent time-lag between the governmental policies and the introduction of the innovations and b) the almost exclusive fixation of the government on the so-called vested interests at the cost of other groups, notably the consumers and the general public (Van Der Loo 1985: 183). These observations are valid for policies concerning new media in the Netherlands. The introduction of Viditel lacked vision: videotex was first preceivedas broadcasting. Then a commercial experiment was initiated without a clear notion of how the project was to be continued and later in spite the experiments results the Viditel-project was confirmed. The policy development surrounding the introduction of videotex was obscure. On the one hand the PTT was asked to take the lead and pave the way for new services on behalf of private enterprise. On the other hand the role of the PTT was so strong and so dominant that business was never given a fair chance. The access of private enterprises to the market of new telecommunications services was nearly impeded. The PTT was mainly preoccupied with maintaining its monopoly of the installation and the maintenance of the information networks against competitors such as cable networks and outsiders such as publishers and other information and service providers.

The introduction of tele-information services, especially for a mass market, is a complex affair, in which a certain level of mutual adjustment among the involved parties is necessary. The product and market development have to be the collective responsibility of the common carriers, service operators and information suppliers. Initially the PTT considered Viditel a specific service which it provided. Later it perceived Viditel mainly as a generic facility that would enable the provision of a wide variety of business and professional information services. Even the PTT was not able to make a succesful videotex system on its own and later realizing that a more cooperative attitude was required in the development of videotex, joined forces with other actors and their systems. The PTT limited its role to 'system integrator' within Viditel and tried to provide more room to the service providers to follow their own marketing strategy.

The expectation that videotex would play an important role in the information supply of the future, led the traditional media of press, broadcasting and telecommunications and the participants in the new electronic media to secure their interests by cooperating in the experiments with new electronic media or by expressing the interest to do so. The new information providers had nothing to loose, and everything to gain; they saw in the videotex service an ideal opportunity for developing new services such as tele-shopping, tele-banking, tele-reservation etc. An example of the new publishers who adopted an offensive strategy is Publimedia (the Dutch 'Yellow Pages - a subsidiary of ITT). It played a stimulating role in the development of videotex in the Netherlands by initiating a branche organization for information providers and by participating in the Infodam- and the Totaalnet-experiments with videotex.

In view of the fact that all actors involved were dependent on the PTT's exclusive infrastructure, the PTT was able to play the role of key actor and compel interorganizational coordination. The power of the PTT is illustrated in the fact, that it (with others) pushed for the harmonization of the various experiments and instigated (with others) the system integrator of Videotex Nederland of which it has a 19.5 % share. The introduction of videotex in the Netherlands was a very complicated game, in which the PTT occupied a central position and in which it held conflicting interests as player, coach and referee all at the same time simultaneously. Being a commonly and legitimately acknowledged key actor in telecommunications policy (as a result of its being a monopolistic state enterprise), it was expected that the PTT - as a coach - should orchestrate and control the diffusion of tele-information services in the Netherlands. As a player, the PTT was subjugated to governmental control and political interference, as well as strategic moves of competitors in the marketplace. Finally the PTT as neutral referee should mediate between the various parties involved in the various experiments with videotex in the Netherlands which in 1989 resulted in a harmonisation of all these experiments in one integrated organization, Videotex Nederland BV. The conflict of interests allowed the PTT to gather strategic information about technology, product-market combinations, market growth and relevant competitors in an experimental setting for a public service for its own corporate purposes.

The PTT was seriously critized from different sides. The publishers accused the PTT of playing publisher with Viditel without giving them the opportunity to participate. The information equipment industry criticized the PTT for imposing restrictive regulations on peripheral apparatus like telephones, printers and terminals. From the very beginning, the press looked upon the new medium with mixed feelings: on the one hand they felt it was a threat to their information monopoly and on the other hand they saw videotex as an instrument for enlarging the readership of newspapers and as such a new source of profit. Their enthusiasm diminished as they realized that due to the slow penetration their investments would not pay off. Industry as a whole stated that because the profits of the PTT were creamed off by the government, there was not enough money left for the necessary investments in developing the telecommunications infrastructure of the future.

The analysis of videotex in the Netherlands leads to the conclusion that the introduction of tele-information services was a political bargaining process between 'pushing and blocking' coalitions of actors, in which the 'established' PTT, with its 'vested interests' in telecommunications policymaking (infrastructure, expertise in tele-infomation services and administrative control) was considered significantly more important than the relative 'outsiders' in the field of telecommunications such as suppliers of telematics, information suppliers, capital investors and private network suppliers. The PTT had important resources at its disposal and had the time to wait for the opportunity to promote the notions of cooperation and integration in the policymaking process. Front stage, the PTT made a few concessions to the others by incorporating its Viditel-service into Videotex Nederland, but backstage the influence of PTT is still substantial: PTT still has an important minority share and has a strong powerbase in network provision. Also PTT's Viditel still had the only operational videotex system on a national scale. So the PTT was in a position to (en)force harmonization among the various experiments, at the moment most conducive to the PTT.

The conditions for relaunching videotex seem to be better than ever. Videotex Nederland can benefit from a nationwide standardized network, a high density of telephones (100% penetration), cable networks (about 86%) and the widespread use of teletekst (58 % of all households). Also the penetration of personal computers (29%) and modems(7%) in households is high; in 1990 approximately 600,000 pc's in private households and 400 000 in business environments, modem penetration 200,000 (Boumans 1990: 45-51). Whether the bundling of the various initiatives will lead to commercial success and a period of prosperity for videotex in the Netherlands, or whether the integration of videotex is merely the integration of a series of failures into a service for which consumer demand is non-existent, remains to be seen. For Videotex Nederland, the system integrator, the message is clear: it's now or never.

References

Algemene Rekenkamer (1989). **Verslag 1988-1989**.The Hague.

Axelrod, R. (1984). **The Evolution of Cooperation**. New York.

Boumans, J.M. (1990). **Elektronische Informatiediensten. Contouren van een Nieuwe Bedrijfstak**. The Hague.

Bouwman, H. & P. Slaa (1990). **Videotex in het MKB. Een Verkenning van Toekomstige Beleidsopties**. The Hague.

Bouwman, H. & P. Slaa (1992). The adoption of Videotex on a Consumermarket. An attempt to predict a Critical Mass. In: **Technolgies de l'Information et Société**.

Burie en Intercai (1987). **Telematica in Nederland**. Nijeholtpade.

Hulsink, W. (1989). Het Machtsspel rond de Invoering van Videotex. In: H. Bouwman & N. Jankowski (eds). **Interactieve Media op Komst**. Amsterdam:.

Hulsink, W. (1991). **Actors Telecommunication Systems and Public Policies. The Diffusion of Videotex in the Netherlands**. Working Paper (Management Report Series No.96) Rotterdam: Erasmus University/Rotterdam School of Management.

Lijphart, A. (1975). **The Politics of Accomodation: Pluralism and Democracy in the Netherlands**. Berkely.

Quelch, J.A. & G.S. Yip (1985). Achieving System Cooperation in Developing the Market for Consumer Videotex. In: R.D. Buzzell (ed). **Marketing in an Electronic Age**. Boston.

Slaa, P. (1987). **Telecommunicatie en Beleid. De Invloed van Technologische Veranderingen in de Telecommunicatie op het Beleid van de Nederlandse Overheid inzake de PTT**. Amsterdam.

Slaa, P. (1989). **Publieksgerichte Videotex in Nederland. Een tussentijds Evaluatie van vijf projecten**. The Hague.

Stuurgroep Viditel (1980). **Eerste Interimrapport ter Begeleiding van de PTT-Praktijkproef met Viewdata**. The Hague.

Stuurgroep Viditel (1981). **Tweede Interimrapport van de Stuurgroep ter begeleiding van de PTT-Praktijkproef met Viewdata**. The Hague.

Stuurgroep Viditel (1982). **Eindrapport van de Stuurgroep ter begeleiding van de PTT-Praktijkproef met Viewdata**. The Hague.

Tweede Kamer (1983-1984). **18368 Interactieve Videotex Viditel.** The Hague.

Van der Loo, H. (1985). Het Wespennest van de Nieuwe Media. In: **Beleid en Maatschappij** 7/8: 175-185.

VNVI (1981). **Zicht op de Toekomst. De Toekomst op Zicht.** Tilburg.

CHAPTER 5

AUSTRIA: AMBITIOUS PLANS...

Michael Latzer
Research Unit for Socio-Economics
Austrian Academy of Sciences
Vienna

Introduction

The story of the Austrian videotex system Bildschirmtext (Btx) is a rare and interesting example of Austrian telecommunications and technology policy.[9] The strenuous efforts made by various public institutions to promote a videotex terminal developed and manufactured in Austria for the world market are rather unusual for a country with such structural characteristics (market size, industrial structure etc.). Another peculiarity occurred when the regulation of videotex led to a long-lasting public conflict over the social risks of the new technical system.
In general, the development of the Austrian videotex system is characterized by a combination of high expectations and low achievements. The result of this combination is a that the limited reputation Btx has acquired in Austria, has been a rather bad one. Nevertheless, among experts, videotex is still considered a promising service configuration for the future, and again and again various efforts have been taken and new strategies have been applied to direct the medium to ultimate success. If this is to occur it will be vital to analyze carefully the shortcomings of the policies applied in the past.[10]

Short National History - Main Players

The importing of the videotex concept to Austria was not originally initiated by the Austrian PTT (Post-, Telephone- and Telegraph-Administration) but by the electronics industry, and Austrian videotex users who participated in the early German Btx trials. In 1977, two years after the first presentation of the original viewdata system in the United Kingdom, ITT-Austria organized the first videotex presentation for TV dealers and, in 1978, the Austrian Consumer Association started in-house videotex tests (Hummel 1982, 2). Next to ITT, Philips Austria was also interested in the early introduction of videotex in Austria. One of their main reasons was that the diffusion of videotex promised a rush for modern TV sets which are needed as videotex screens.
In the late 1970s, Prof. Maurer and his Institute for Information Processing (IIG) in Graz, established themselves as a central driving force for videotex development in Austria. Later on, an informal partnership between PTT and IIG was formed, not least because of the capacity problems of the PTT.[11] The interests of the IIG were focussed on the development of the videotex terminal (decoder), to be connected to a TV set, and on telesoftware.
To sum up, the main interests of the early lobbyists for videotex in Austria were in the equipment rather than the service side.

A stronger videotex involvement of the PTT was promoted by Dr. Heinrich Übleis,

---

[9] For background information about Austrian telecommunications see Bauer/Latzer 1988.

[10] In essence, this paper is based on interviews with experts. For comments on earlier drafts I want to thank Prof. M. Elton, DI R. Michalke, Dkfm H. Neßler and Dr. T. Vedel.

[11] As part of the General Administration, the PTT is bound to restrictive administrative rules and therefore cannot react to changes in the markets as flexibly as sometimes needed.

H. Bouwman and M. Christoffersen (eds.), Relaunching Videotex, 53–67.
© 1992 *Kluwer Academic Publishers.*

who became General Director of the PTT in 1978 and, generally, emphasised more strongly than his predecessors the new technological developments in the telecommunications branch of the PTT.

Major steps towards an Austrian videotex service were taken in 1979 by
- a 'Btx Administrative Agreement' in which the Ministry of Transport (the general directorate of the PTT forms a division of this ministry) and the Ministry for Science and Research who stipulated their support for a videotex service and by
- the establishment of the 'Videotex-type Systems' working group, chaired by Prof. Maurer, in the Austrian Computer Society.[12]

Finally, in March 1981, the PTT started Btx pilot trials for up to 300 subscribers, which were originally scheduled to last until the middle of 1982. The regular service was intended to commence in 1983. (Hummel 1982, 7)
A critical assessment[13] of the videotex pilot phase by a research group commissioned by the Ministry of Science and Research was not considered by the decision makers. The plan for the introduction of the service had already been determined before the results of the research project became available.

Nevertheless, the introduction was still delayed for more than two years. The main non-technical reason for this delay was the public controversy about the regulation of the possible social impact of videotex. Finally, in 1985, a compromise was found and Btx was introduced as a regular service. However, the painful emergence of the service was followed by major acceptance problems which forced the PTT to change the introduction strategy and conceptual definition of Btx. Acceptance problems not only occurred for Btx in Austria but also internationally for the Austrian-developed Btx terminal, and finally led to the total breakdown of an ambitious national technology policy project.

---

[12] The working group existed until 1990 and was then renamed 'Hypermedia Systems'.

[13] See Hummel 1982, Pilz/Werthner 1982; In their analysis, the authors outlined social risks and criticized the Austrian videotex plans from an economic point of view.

Table 1. Main Players in the Austrian Videotex Development - Overview

Player	Brief Characterization
PTT (Ministry of Transport)	System operator; interest: better capacity use of the telephone network
Inst. f. Information Processing (IIG), Prof. Maurer	Graz branch of the Austrian Computer Society; R&D of MUPID and telesoftware;
Ministry of Science & Research	Subsidized the development of MUPID; finances partly the IIG; technology and science policy;
Austrian Consumer Federation	Btx pioneer; first Btx inhouse network; lobbyist for Btx law;
Electronic Industry: ITT, Philips, Siemens	Btx equipment suppliers; early Btx lobbyists;
Social Partnership	Informal body; preparation of legislative regulation of Btx
Government of Styria	Financial aid for Motronic; interest: creation of jobs and job-security
Motronic	Production of MUPID;
Mupid Computer GesmbH. (MCG)	Marketing and development of MUPID;
Infonova	Successor organisation of MCG
MUPID Computer Club Austria (MCCA)	Association of Btx-users; since 1989: Microcomputer Club
Btx Anbietervereinigung (BAV)	Association of Btx-information providers; dissolved in 1989;

General Description of the Current Videotex Situation

THE CHOISE OF TECHNOLOGY AND ARCHITECTURE

A system decision of extreme significance had already been made in the late 1970s. Without any controversial discussion (Hummel 1982), the service configuration of the British videotex system, Prestel, was chosen for the Austrian pilot trials. The decision in favor of the the hierarchical architecture of Prestel was supposedly made because the German and Swiss PTTs had at this point already chosen the same system. However, the Prestel standard was only used until June 1985. Then the system was converted to the CEPT C2 standard[14], a decision which was especially costly for the PTT because, in addition to the new central units plus software, all the rented Btx-terminals had to be exchanged without charge and the already existing information pages had to be converted as well.[15]

The configuration of the Austrian videotex system is shown in Figure 1. The Btx-system comprises one central control center in Vienna and five regional Btx centers in Innsbruck, Graz, Salzburg, Klagenfurt and Vienna, operated by the PTT and equipped with British GEC computers and software adapted by Siemens Austria. The control center is connected to the regional centers by direct data lines, the third-party external computers to the regional centers via the packet-switched network.

The problem with the hierarchical architecture is that every connection has to go through the Btx center of the PTT. This is why the Btx centers become the bottleneck of the system. A service with 100,000 subscribers, as already forcasted for 1987 by the PTT in 1980, could not be handled even by the central computer currently used (the maximum would be around 50,000 subscribers). Furthermore, only 200 to 300 subscribers[16] can use Btx at the same time. None of these problems exist in the French system, which uses no central computers but so-called videotex access points. Another disadvantage of the hierarchical architecture compared with the access point structure is the higher costs for information providers who intend to connect their computers to the videotex system.

On the terminal side, it was decided to develop a graphics decoder (without integrated screen) called MUPID (Multipurpose Universally Programmable Intelligent Decoder). MUPID, based on a 8 bit processor (Z80A), was developed at the IIG (Prof. H. Maurer) and manufactured by the small Austrian company Motronic. MUPID is to be attached to a modern TV set[17], and in addition, MUPID boards for Personal Computers (PCs) were developed.

Normally, Btx is accessed via 1200/75-baud modems. In 1990, technical trials of "fast Btx" using 1200/1200-baud and 2400/2400-baud modems were started and now offer the possibility of using more common modems. The current problem with fast Btx is that, instead of charging only local telephone tariffs regardless of distance, both local and long distance telephone tariffs are charged.[18]

An important, characteristic feature of the Austrian videotex system is the extensive use of telesoftware which makes it more powerful than many other sytems. A software library can be stored in the Btx centers, downloaded by the subscriber when needed and then used without a telephone connection.

---

[14] CEPT, the interest-organisation of the European PTTs, basically distinguishes between three videotex standards: CEPT 1 (Bildschirmtext), CEPT 2 (Teletel) and CEPT 3 (Prestel). Within CEPT 1, three specifications C0, C1 and C2 were defined. Whereas Germany and Switzerland opted for CEPT1 C0, only Austria chose the more sophisticated (vector graphics) and expensive C2 standard.

[15] The total costs were about 110 million Austrian Schilling (7.6 mio ECU) (Wochenpresse, 1985-6-11). Austrian Schilling are converted to ECU throughout this paper using average annual exchange rates. (1 ECU = 14.56 ATS)

[16] Per regional Btx center.

[17] A SCART-socket is needed.

[18] However, the PTT intends to charge only local telephone tariffs starting the end of 1991.

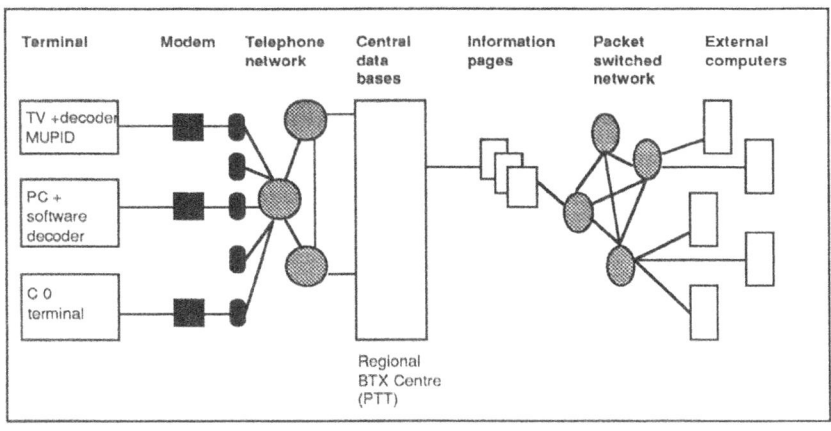

Figure 1: Configuration BTX-Austria

CONCEPTUAL DEFINITION AND DEVELOPMENT STRATEGY

The conceptual definition of Btx in Austria has been exposed to major shifts in emphasis over the last ten years. Originally, in the late seventies and early eighties, Btx was seen as a new medium, a mass service, which would be widely accepted by private households as well as by business users. Btx was described as network of microcomputers, using 'intelligent' terminals with a high graphics standard (the Austrian innovation MUPID), and an interactive mass service, connecting private households with companies. Around 1986, after years of bitter acceptance problems, especially with private households, the promotion strategy of the PTT was redirected more towards business users, e.g., to form closed user groups and to use Btx as an in-house network. In the late 1980s, the hopes and corresponding promotion of Btx as a future mass service returned. The new optimism was based on the fact that various steps had been taken, especially since 1986, to enhance the attractiveness of Btx:

- The use of acoustic couplers was made possible, public Btx terminals installed, and Btx/telex-, Btx/eunet-usenet- and Btx/pager- gateways introduced.
- The PTT started a massive promotion campaign for Btx in 1986.[19]
- Hoping to solve the often cited chicken-and-egg problem[20] the PTT extended the list of services offered by the electronic telephone book[21] in 1988.
- Since 1989, the Austrian system has been interconnected to the videotex systems of Germany, Switzerland and Luxembourg.

---

[19] About 4 million Austrian Schilling (0.27 mio ECU) were spent in 1986. The campaign was criticized for raising too high expecations which could not be fulfilled and, therefore, finally led to correspondingly deep disappointment. However, the reactions by the potential users were rather limited.

[20] No users as long as no information providers and no information providers as long as there is no critical mass of users.

[21] This service was a major part of the successful French introduction strategy - but only in combination with free terminals for the subscribers.

- The terminal market was liberalised. Until 1989, MUPID was used in the Austrian Btx system almost exclusively (this protectionism is quite usual in the international tele-communications sector). Since 1989, software decoder[22] for industry-standard personal computers have offered a cheaper way to access Btx and also widened the potential group of users to about 300,000 PC owners in Austria .

To sum up, the Austrian PTT is liberalizing the terminal market, enhancing the attractiveness of the service side, and partially adapting to the successful French teletel system. With nearly all the changes in the Btx strategy, the Austrian PTT more-or-less followed the changing strategies of the German Bundespost (see Schneider 1988). The diffusion strategy of the PTT is now, after years of concentration on business users, again more directed towards private households in order to make Btx a mass service. One major problem of all these efforts is the non-existent cooperation with information providers.

DIFFUSION AND USAGE OF BTX IN AUSTRIA

The introduction of Btx in Austria was accompanied by extraordinarily high expectations which could not be met by actual developments (see table 2). Nevertheless, these expectations formed the basis for speculation about the impact of Btx and hence led to a public controversy between the Social Partners. The example shows that exaggerated forecasts of the Btx lobbyists did not function as self-fulfilling prophesies and promote Btx, but instead as self-defeating prophesies, which finally turned against their spiritual fathers and helped impede the diffusion of Btx. The grand promises that were made especially by PTT and IIG might also have led to excessive expectations on the part of potential users and, hence, to greater disappointment by subscribers than objectively would have been justified by the actual service quality.

A major technical obstacle to widespread diffusion of Btx was obviously underestimated in all prognoses and plans. It is a peculiarity of the Austrian system that only one half of the telephone main lines are single lines and the other half, in the biggest cities, Vienna and Graz, even up to 75 per cent[23] in the mid-1980s, are so-called "party lines" where up to four sub-scribers share one line. That means, if one of the four is on the phone, the rest cannot make or receive calls. The subscribers of party lines are consequently not allowed to join Btx. Another technical obstacle for the early diffusion of Btx was that in order to use MUPID the TV set needs a SCART socket which was not widespread at the time of the introduction of the Btx service.

In table 2 some forecasts about the diffusion of Btx and MUPID are contrasted with the actual developments. Interestingly, in Germany quite similar incorrect prognoses have been made. In absolute numbers they differ, corresponding to the difference in population, approximately by the factor 10 (see Schneider 1988). The result is that Austrian and German prognoses correlate with each other but not with actual developments.

---

[22] The PTT sponsored the development of public-domain software decoders.

[23] The high proportion of party lines is, in essence, a legacy of hasty reconstruction after the second world war.

Table 2. Btx Forecasts and Btx Reality in Austria

Forecaster (year of forecast)	Prognosis	Reality
Minister for Transport (1980)	1988: 100,000 subscribers	8,856 sub.
PTT (1984)	1987: 100,000 subscribers	7,893 sub.
Motronic (1984)	1984-86: 190,500 sold MUPIDs (Prestel standard) in Europe[1] starting 1985: 200,000 per year;	until 1985: 6,000; then C2-Mupid hardly exported;
Diebold (1985)	1990: 80,000 subscribers[2]	9,887 sub.
PTT (1986)	end of 1986: 9,000 sub.	6,228 sub.
PTT (1986)	end of 1987: 16,000 sub.	8,323 sub.
1. 0.25% of all TV-households in Europe plus 50,000 to FRG; 2. average annual growth of 80%; Sources: Hummel 1982; APA-Medien, 1985-8-9,p.6;p.7; Pilz/Werthner 1982,46; APA-Medien, 1986-8-29, PTT;		

The actual development of Btx is illustrated in table 3. In August 1991, there were 12,940 subscribers, this is a distribution of 1.7 Btx subscribers out of 1,000 inhabitants (4 per 1,000 telephone subscribers). However, 867 of those terminals (close to 8 per cent) are PTT terminals used for internal and test purposes. While the number of users was rising, the number of information providers has been drastically declining, from 762 in 1986 to 389 in August 1991. Moreover, the number of third-party external computers connected to the regional Btx centers of the PTT fell to 51.

More than a third (37.8 per cent) of Btx subscribers are located in Vienna (19.5 per cent of the country's population). In 1990, 250 closed user groups, mostly professional users, existed.

In August 1991, 58,160 information pages, about 102,537 KBytes ('blocks') of information, were in use. The number has been declining since 1989 (see table 4).

The average use time of Btx has also declined over the last few years, from 8 hours per month and user in 1986 to 6 hours per month and user in 1989. This can be explained partly by the fact that the users became more efficient, but it could also mean that the attraction of the 'new' medium is decreasing. In 1989, the users spent about 40 per cent of the usage time in the databases of third-party external computers.

According to estimates by the PTT, only 5 per cent of the Btx users are private households, compared to 80 per cent professional and 15 per cent semi-professional.[24] The diffusion of Btx by economic sectors for 1986 and 1990 is shown in table 5. According to this data, the use of Btx is highest in the banking and insurance sector. A study by Hummel/Kissinger (1986) suggests that Btx is in general largely unknown in the Austrian trade sector, and hence any effect of its impact could hardly be analysed due to the low diffusion of the medium.

---

[24] Lawyers, architects etc., who use Btx for business as well as for private purposes.

Table 3. Diffusion of Btx in Austria

Date	User	Inf. Prov.	Ext. Comp.	PTT-Term.	Total	Growth	Growth Rate
av. 81	75	67	0	1	143		
av. 82	135	123	0	0	258	+115	
av. 83	161	277	0	0	438	+180	
av. 84	872	351	8	65	1296	+858	
av. 85	2557	685	24	102	3368	+2072	+16.1
av. 86	3819	762	40	165	4786	+1418	+42.1
av. 87	6820	755	43	275	7893	+3107	+64.9
av. 88	7743	661	50	402	8856	+963	+12.2
av. 89	8008	595	55	694	9352	+496	+5.6
av. 90	9445	491	56	852	10485	+1493	+16.0
Aug. 91	11633	389	51	867	12940	+2095	+19.3

(source: PTT)

Table 4. Number of Btx-Pages and Information-"Blocks" (KBytes)

Date	Pages	Growth	Blocks (1 Kbyte)	Growth	Info. Prov.	Blocks Per Inf. Prov.
av. 85	49136					
av. 86	70510	+21374				
av. 87	72042	+1532	99649		755	132
av. 88	77156	+5114	107132	+7483	661	162
av. 89	77392	+236	109110	+1978	595	183
av. 90	73043	-4349	109012	-98	491	222
Aug. 91	58160	-14883	102537	-6475	389	264

(source: PTT)

Table 5. Diffusion of Btx by Economic Sectors, 1986, 1989:

1986 : (estimated)	banks		30%
	trade and industry		27%
	public services		5%
	semiprofessionals and private users		20%
1989:	public services	(incl. ministries, hospitals, PTT etc)	9.3%
	educational sector		4.8%
	medicine		1.2%
	industry		3.1%
	trade		20.6%
	bank, insurance		19.3%
	traffic		3.0%
	lawyers		7.0%
	others		31.7%

(source: Postrundschau 6/1986, 13; PTT)

TARIFF POLICY AND EQUIPMENT COSTS

In Austria, as opposed to the French introduction strategy, terminals are not free of charge, but can be rented from the PTT for a subsidized fee. Since 1989, so-called rent-and-buy schemes have been offered, in which, after a 48-month rental period, the unit becomes the subscriber's property. Under the pressure of low Btx acceptance the costs of using Btx decreased over the years and the tariff structure became more flexible to the needs of different types of users, e.g., the possibility was created of buying a user number without having a personal terminal.

Information providers can charge extra money for the use of their information pages.[25] The charges are collected by the PTT with a separate Btx bill and then transfered to the information provider. The PTT keeps four per cent of the collected charges. A disadvantage of this billing system compared to a time-sensitive system as in France is that the user has to decide to pay the full price for a page without knowing if the contents are of any relevance to him. In the French system the user has the chance of paying only until he decides that the contents are of no interest to him.

The main components of Btx costs for users and information subscribers are summarized in table 6.

---

[25] Up to 99 Austrian Schilling for one page and up to 1,000 Austrian Schilling for a block of information pages.

Table 6. Btx Costs for Users and Information Providers, 1991

Charges for Btx Users:
•   installation fee: 400 ATS, 27.5 ECU;
•   no standing charges;
•   local telephone rates (40 ATS/h, 2.8 ECU/h) as time charges for the use of Btx;
•   the price for chargeable pages is set by the information providers and collected by the PTT with a separate Btx bill; 96% of the revenues are transfered to the infor^_mation providers;
•   subscribers without a personal Btx terminal are charged 400 ATS (27.5 ECU; non-recurring) for a Btx user number;
Equipment Costs:
•   since 1989: PTT has been offering rent-and-buy schemes = possibility to buy the Btx equipment with 48 monthly rates:
+   MUPID: 130 ATS/month, 8.9 ECU/m (price: ATS 6,240; 429 ECU)
+   Btx modem: 70 ATS/month, 4.8 ECU/m (price: ATS 3,360; 231 ECU)
+   or acoustics coupler: 90 ATS/month, 6.2 ECU (price: ATS 4,320; 297 ECU)
•   MUPID extension board for PCs: 15,000 - 20,000 ATS; 1030-1374 ECU;
•   Decodix (decoder-software for DOS-systems): public-domain software; 300 ATS, 20.6 ECU for cabel and manual;
Charges for Btx Information Providers:
•   "cover-page" (monthly): 500 ATS, 34.3 ECU;
•   info. block (1kB) (monthly): 9 ATS, 0.6 ECU;
•   closed user group (monthly): 500 ATS, 34.3 ECU;
•   connection of external computer (monthly): 1,750 ATS, 120.2 ECU;

Policy Analysis

THE CONFLICT OVER THE REGULATION OF BTX

Within the Social Partners, differing views existed on the social risks of the widespread use of Btx. The Trade Unions and the Chamber of Workers together with the Austrian Consumer Association intervened against the scheduled start of the regular Btx service in 1984, which in consequence had to be postponed for over a year. The major issues in the conflict concerned data privacy (user profiles), consumer protection (the right to return merchandise), uncontrollable lay offs and the rise of electronic homeworking (labor legislative problems especially for women) due to the expected widespread use of Btx in various branches of the economy.

In essence, the Chamber of Workers demanded a Btx law (comparable to the German Btx law) and the Chamber of Commerce, and not only the conservative People's Party, but also representatives of the Social Democratic Party, were against any restrictions which could have possibly limited the use of Btx. (Neue Zeit, 1984-4-17; Kurier, 1984-6-13)

Finally, in 1985, a compromise was reached and the regular Btx service started. The results of the negotiations between the Social Partners were integrated in the Btx user guidelines of the PTT and, furthermore, the Social Partners agreed to start work on a future Btx law. Concerning the data privacy risks, two million Austrian Schilling (0.14 mio ECU) were spent to develop and implement a software solution allowing anonymous entry to Btx. However, the user can only remain anonymous as long as he/she uses services that are free of charge. Real anonymity, as in the French kiosque system, remained impossible in Austria due to the technical specifications of the telephone network.

Interestingly, in 1990, the user guidelines of the PTT are still the only legal basis for the regular Btx service. The Btx law never passed draft stage.[26] The main reason for this delay seems to have been the low acceptance of Btx, which reduced the relency of the discussion about the possible risks in Austria.[27]

To sum up, the whole conflict over the regulation of Btx was to a great extent based on, and triggered by, wrong assumptions and forecasts of its importance in economic and private life. Nevertheless, these circumstances led not only to a painful delay in the introduction of the service but also seemed to harm the reputation of Btx and hence its diffusion.

In December 1988, the Administrative Tribunal stated that the relation between PTT and Btx-subscribers is regulated by private law and not by public law as for instance the telephone service. That also means that as private service the telecommunications monopoly of the PTT cannnot be applied for Btx and other value added services.
The trigger for all these interesting legal insights was a so-called "sex corner" in the Btx-newspaper which also includes a lexicon about perverted love-practices which made the PTT disconnect the information provider of this service. Hence, the provider went to court... (EDV&Recht 1/89,33-35; profil 22.May 1989, 32).
Controversy about the "sex corner", Btx-newspaper, including a lexicon of perverted love practices; PTT turned this info. provider out; inf.prov. went to the Administrative tribunal (supreme court); the Administrative tribunal ....)

TECHNOLOGY AND INDUSTRIAL POLICY EFFORTS

The Austrian innovation MUPID, which was basically initiated by Prof. Maurer (IIG), University Professor in Graz, and Dr. Übleis, the PTT's Director General, is a rare and remarkable Austrian example of explicit, concerted technology and industrial policiy in the international telecommunications sector, with various state institutions, and either publicly-owned or publicly-influenced institutions involved.
In the late 1970s, the trigger for the national development of a personal videotex terminal was the encouragement of the PTT General Director and his definite promise to procure 500 units. Prof. Maurer, who until then had worked exclusively theoretically, accepted the challenge. Übleis proposed Motronic, a small company in Styria which had already worked for the PTT, as industrial partner its the manufacturing.
Major financial support for research and development at Maurer's institute (IIG) in Graz was provided by the Federal Ministry for Science and Research, which spent about 8 million Austrian Schilling (0.55 mio ECU) until 1985 (Wirtschaftsnachrichten, 15.6.1985). The Austrian Computer Society (OCG) was chosen as mediator between the Ministry and the IIG.[28] The involvement of the Ministry was, inter alia, justified by the argument that several hundred jobs would be secured or created by the production of MUPIDs in Austria.[29]

---

[26] The draft Btx law, a transformation of the PTT user guidelines into law, also includes the establishment of a Btx committee at the Ministry of Transport consisting of representatives of the Social Partners. (Medien und Recht 1/1987, 3).

[27] For a discussion of legal aspects of Btx see Duschanek 1989.

[28] The IIG (Institutes for Information Processing), which in 1990 comprised eight institutes, is officially the Graz branch of the Austrian Computer Society but holds special privileges: It has been granted the status of an independent society with its own president (currently Prof. Maurer) and obtains subsidies directly from the Ministry for Science and Research.

[29] Motronic had about 200 employees.

In the autumn of 1982, the PTT procured the first 250 MUPIDs. Until 1989, MUPID was as far as possible protected from competition in Austria, basically by the licensing policy of the PTT. Furthermore, the PTT procured many more MUPIDs than were really needed. In 1988, this led to the situation that the PTT had over 5000 MUPIDs[30] in stock, which even provoked criticism by the Austrian Audit Office (Rechnungshof) (Standard, 9.10.1989, 1989-4-24). The reason for such high stocks was a mixture of shipwrecked industrial policy and wrong prognoses concerning the diffusion of Btx in Austria. In 1990, the PTT probably still has 4,000 MUPIDs in stock and an important question is how these circumstances are influencing the further Btx strategy of the PTT.

From 1981 to 1989 the PTT investment in Btx amounted to 731 million Austrian Schilling (50.2 mio ECU) (see table 7), about 87,000 Austrian Schilling (5,975 ECU) per user. Altogether, in 1989, the expenditures of the PTT on Btx had already exceeded one billion Austrian Schilling (68.7 mio ECU). The revenues from the Btx service were, for example, in 1988 about 75 million Austrian Schilling (5.2 mio ECU)[31] (Standard, 1989-4-24).

The support for the MUPID-project was not limited to the Ministries of Transport and of Science & Research. In 1983, the Federal Ministry of Education also suddenly started to help the diffusion of Btx and MUPID. The Ministry bought 231 MUPIDs for schools, investing about 10 million Austrian Schilling (0.7 mio ECU) in equipment (Profil, 4/1986, S.47). This led to complaints from teachers who did not see a need for Btx in schools and furthermore criticized the complete lack of plans or suggestions on how to integrate Btx into their daily work (AHS-aktuell 37/1986). Moreover, problems of financing an additional telephone connection hampered the use of Btx in schools.

Further support for the MUPID technology and industrial policy project came from the Austrian nationalized industries and the Bundesland Styria, where the Austrian company Motronic, which had been founded in 1970, was located.

Even in the early 1980s it had become obvious that Motronic was overburdened by an industrial project of this size. In order to take over the (inter)national marketing and development of MUPID, the MUPID-Computer GmbH (MCG) was founded in 1983 by Motronic, Siemens Austria[32] and the nationalized companies VOEST and Elin, with each partner holding a 25 per cent share. In the same year, a subsidiary, MCG-Deutschland was established in Munich (Börsen-Kurier, 1983-9-15). But the problems did not end. In 1983, the continuation of Motronic was in danger when two partners who basically had built up Motronic left the company because of controversies within the management. In this emergency situation, the Styrian Bundesland decided to take over the majority (51 per cent) of Motronic.[33] Two years later, in 1985, Motronic again accumulated high debts which were written off (A3-volt 12/85, 6). The Styrian share in Motronic was reduced to 39 per cent, the rest was bought by the Austrian Assmann Industrial Group.

However, until 1985, the world-market situation seemed to be most favorable for the Austrian MUPID project. In 1983 and 1984, the MCG had experienced its successful years. In

---

[30] With a value of about 70 million Austrian Schilling (4.8 mio ECU).

[31] Excluding telephone revenues derived from Btx.

[32] At that time still partly owned by the state.

[33] Motronic had already been a "classical" case for economic funding by the Government of Styria before the production of MUPID started. Now Styria initiated and financed a newly formed association which hence took over the majority of Motronic. This construction was chosen in order to be more flexible than within the bureaucracy and also because with this construction the case remained in the field of responsibility of the Wirtschaftsförderungs Department which was dominated by the ruling conservative People's Party and would have otherwise been "lost" to the responsibility of the Finance Department, which in Styria is traditionally dominated by the Social Democratic Party.

1985, it still had a turnover of about 100 million Austrian Schilling (6.9 mio ECU) and an export share[34] of 15 per cent (Wiener Zeitung, 1986-1-17). A 40 per cent increase in turnover was expected for 1986 and the Minister for Science and Research praised the successful innovation and proudly presented the second generation of MUPID[35] to the public as the most modern videotex computer (Arbeiterzeitung, 1985-6-15).

The situation changed quickly when the MCG started to run up heavy losses.[36] From 1983 to 1985, the MCG had a turnover of 260 million Austrian Schilling (17.9 mio ECU) with 90 million Austrian Schilling losses (6.2 mio ECU) (trend 1989/10, p.224). The export opportunities especially to Germany had been greatly overestimated (see table 2). Furthermore, Siemens announced its plans to leave the MCG mainly because of conflicting interests within the company.[37] Negotiations with Philips and ITT, which were in principle interested in taking over the Siemens share, failed, and finally Motronic bought the MCG for a symbolic price of 1 Austrian Schilling.[38] In return, Assmann had to guarantee 160 jobs for Motronic over the coming five years.

However, the downward trend of the MUPID-industry continued. In 1989, Assmann had losses of over 100 million Austrian Schilling (6.9 mio ECU) and asked the Federal and Bundesland governments for financial support. Finally the production of MUPID had to be stopped, the MCG was dissolved and, in 1990, parts of Motronic were also sold. At the end of 1989, three former employees of the MCG founded the Infonova company which took over the marketing and maintenance of MUPID.

As early as 1986, Prof Maurer and the IIG had more or less given up MUPID (but not Btx), when Assmann decided not to produce the third generation of MUPID[39], which at that time had already been finished in the lab. Hence, Maurer and his group concentrated on the development of Btx software decoders for PCs.

With the liberalization of the terminal market by the PTT and the ability to use C0 terminals, the MUPID project is dead. Paradoxically, this could at the same time be the basis for the survival of the Btx service.

---

[34] MUPID was for instance exported to the Netherlands, Sweden, Norway, Germany, Switzerland, United Kingdom, South Africa and Australia (Presse, 1985-6-25).

[35] For the second generation of MUPID, the CEPT C2 standard instead of Prestel was applied.

[36] Even in 1983, Prof. Maurer had stated, that in order to manufacture without financial losses, 50,000 MUPID chips would have to be sold. In 1985, only 15,000 were sold (a3 volt 12/1985, p.7).

[37] One reason might have been the decision of the MCG in 1985 that AMI, a VOEST subsidiary, and not Siemens would develop and manufacture the highly integrated chip for the next MUPID generation.

[38] In 1986, the subsidiary MCG-Deutschland was dissolved.

[39] In this compact MUPID, modem and screen are integrated, and a customized LSI chip (see footnote 29) is used.

Table 7. Investment of the Austrian PTT in Videotex

- Until		
1982:	31.8 mio ATS	2.2 mio ECU
- 1983:	53.8 mio ATS	3.7 mio ECU
- 1984:	137.0 mio ATS	9.4 mio ECU
- 1985:	269.0 mio ATS	18.5 mio ECU
- 1986:	86.0 mio ATS	5.9 mio ECU
- 1987:	53.0 mio ATS	3.6 mio ECU
- 1988:	70.6 mio ATS	4.8 mio ECU
- 1989:	<u>30.0 mio ATS</u>	<u>2.1 mio ECU</u>
Total	**731.2 mio ATS**	**50.2 mio ECU**

Conclusions

Videotex in Austria is made up of two closely interlinked technology policy projects, one on the service side, and one on the terminal side. Both, however, were of little success. Btx has about 12,000 subscribers and would probably need around 50,000 to break even; the Austrian innovation MUPID failed to succeed in the world market and, after nine years of strenuous technology and industrial policy efforts, its production was finally stopped in 1989.[40] A positive side-effect might be that cheaper access to Btx might now enhance its diffusion in Austria.

National support for Btx and MUPID not only came from the PTT and the Federal Ministry of Pubic Economy and Transport, but also from the Federal Ministry of Sciences and Research, the Federal Ministry of Education and Sports, the Government of the Styrian Bundesland and the nationalized industries, which have altogether already invested more than a billion Austrian Schilling (68.7 mio ECU) in the technology and industrial policy project. The PTT is still the unchallenged main player in the Austrian videotex sector; information providers play a minor, even negligible role.

Following many interviews with experts, various reasons can be identified which contributed to the failure of the Austrian videotex endeavours:
* Service side: hierarchical architecture; poor marketing; exclusiveness of MUPID terminals; missing preconditions in the technical environment[41]; the neglect of the social dimension of the service; wrong and over optimistic prognoses for the diffusion and use of the service; public conflict over the regulation of Btx; cooperation problems between information providers and the PTT.
* MUPID project: expensive technical standards; small domestic market; international protectionism in the telecommunications sector; poor marketing; cooperation problems between university and industry; coordination problems between the various supporters of MUPID.

To sum up, the Austrian case is an example of the, theoretically already well-known experience, that even brilliant technical ideas are no guarantee of economic success, and, therefore, technology policy decisions should only be taken on the basis of a thorough strategic business plan which assesses the economic and social risks and chances of a project. Moreover, the Austrian example underlines the necessity of carefully defining the interface and different

---

[40] Altogether, about 10,000 Prestel-MUPIDs (1982-1985; 60 per cent exported), and 14,000 CEPT C2 MUPIDs (1985-1989, hardly exported) were produced.

[41] Single telephone lines; SCART-sockets;

tasks of cooperation between university and industry. In general, there remains the question of whether a small and open economy like Austria, without a multinational company, should try to develop, produce and manufacture high-tech equipment for the world market on its own.

On the service side, the PTT more or less followed the changing German strategies, from the definition of Btx as a mass service until 1985, to the concentration on business users from 1986 to 1988, and carefully back to the mass market since 1989. Once again as in Germany, the attractiveness of the service was improved, e.g., by offering "fast" Btx, the electronic telephone book and by liberalizing the equipment market.[42] Whether these efforts will have a strong positive effect on the development of the service, or whether these are just "cosmetic operations on a corpse"[43], remains to be seen.

References

Bauer, Johannes & Latzer, Michael (1988). Telecommunications in Austria, pp. 53-87. In Foreman-Peck, James & Müller, Jürgen (eds): **European Telecommunication Organisations, Baden-Baden.**

Duschanek, Alfred (1989). Rechtliche Grundfragen des Bildschirmtext-Betriebes, pp. 219-249. In Korinek, Karl & Stampfl-Blaha, Elisabeth : **Beiträge zum Telekommunikationsrecht, Wien.**

Hummel, R., Eichinger, M. & Haberl, G. (1982). **Begleitstudie zur Einführung von Bildschirmtext in Österreich.** Forschungsbericht, Wien.

Hummel, Roman & Kissinger, Robert (1986). **Auswirkungen von Bildschirmtext auf die Struktur des Warenhandels in Österreich.** Forschungsbericht, Wien.

OECD (1988). New Telecommunications Services. Videotex Development Strategies, Paris.

Pilz, Peter & Werthner, Hannes (1982). **Ökonomische Bedeutung der Neuen Medien in Österreich.** Forschungsbericht, Wien.

Schneider, Volker (1988). **Technikentwicklung zwischen Politik und Markt: Der Fall Bildschirmtext,** Frankfurt.

Periodica:

A3-Volt   Postrundschau   Kurier
Arbeiterzeitung   Presse   Medien und Recht
Börsen-Kurier   Profil   Trend
Die Zeit   Standard   Wochenpresse

---

[42] There is still some reluctance to change the whole architecture of the system, which in the current configuration could be a bottleneck for a real mass service Btx.

[43] Statement by the president of the Austrian Consumer Federation; (Wochenpresse, 1990-4-13, .p.47)

CHAPTER 6

SWITZERLAND: A MODEST SUCCESS IN TINY PRAGMATIC STEPS

Heinz Bonfadelli
Seminar für Publizistikwissenschaft der Universität Zürich

Introduction

Switzerland is a small, multicultural state in the center of Europe. At the end of 1989 the population was 6.7 million living in 2.5 million households. More than one seventh of them were foreigners. Due to the topographical situation of the country, this population is unevenly distributed. Half of the population live in suburban or metropolitan areas. This contrasts considerably with the thinly populated regions and peripheries. The country is culturally segmented in four different language groups: German, 74%, French 20%, Italian 5%, Romansch 1%. These figures correspond to the year 1980.

As in other European countries technological innovations, international developments, market forces and changes in media politics have altered especially the electronic media aswell as the telecommunication system in Switzerland since the 1980s (Bonfadelli & Hättenschwiler 1989). The following trends in the German part of Switzerland are based on representative survey-data gathered by the audience department of the Swiss Broadcasting Corporation (SRG Forschungsdienst 1991): due to topographical reasons the installation of electronic *cable systems* in television households increased from 30% in 1980 to 70% in 1990; the average number of *television channels* available has doubled from 6 to 13. As a result, more than 60% of the *viewing time* is now spent watching foreign television programmes. In the same period the diffusion of videorecorders has increased from 3% to 43%.

Experiments with over-the-air *teletext* began in 1981. This service, a joint project by the Swiss Broadcasting Corporation and the Swiss Newspapers Association, was introduced to the public in 1984. Today 38% of the television sets are equiped with a teletext decoder. This service is used frequently by one fourth of the television audience.

Switzerland has a highly developed *telecommunication system* with 3.9 million telephone main switchboards. Its complete digitization is the most prestigeous project at the moment: by 1992, 85% to 90% of the existing 1,000 telephone exchanges will be digitized. In addition, the PTT, *the Swiss Postal Service*, launched several new telecommunication services that have been highly successful: *cellular radio 'Natel C'* rapidly reached 125,047 subscribers and 97,609 PTT or private telefax devices already had been installed at the end of 1990. Technical experiments with telephone-based *videotex* started in November 1979, followed by a two-year public testing-phase between September 1984 and end of 1985. The public Swiss videotex system started offically on January 1. 1987. Since then the numbers of information-providers and information-users of Swiss videotex has increased at a slow but continuous rate: today, Swiss videotex is a modest success, achieved in tiny, pragmatic steps.

In contrast to foreign solutions, the PTT wishes to keep its unity of non-electronic postal and telecommunication services in the future. But a modest liberalization of its monopolistic position and a concentration on basic services will be inevitable as the public and parliamentary debates on the new Telecommunication Law show. This will probably take place next year. In a first step, the PTT have opened user end-devices like phones, modems, telefax etc. to the free market. Nevertheless, these devices have to be technically approved by the PTT. In a next step PTT will give a chance to privately operated *audiotex services*.

H. Bouwman and M. Christoffersen (eds.), Relaunching Videotex, 69–84.
© 1992 *Kluwer Academic Publishers.*

Current Situation of Videotex in Switzerland

In 1987, after a seven year long testing-phase, the Federal Government of Switzerland decided to permit public videotex service in Switzerland. Based on a Federal Videotex Decree, the national PTT was authorized to market the system to subscribers and providers of information. Despite early optimism, the task of converting a technical system into a useful public service has not yet been solved satisfactory. The diffusion of videotex has been rather slow until the last few years when the subscriber rate almost doubled each year and increased from 14,000 at the end of 1988 to 75,000 in July 1991. This represents the second highest videotex density in Europe after France: one videotex subscriber per 90 inhabitants or per 33 households.

SYSTEM STRUCTURE

Based on the CEPT standard, *Alcatel STR* installed for the Swiss PTT the first multilanguage videotex main computer in Berne at the end of 1983. Its main functions are: 1. internal data base, 2. inventories (topics, information providers, subscribers), 3. electronic mail, 4. identification of videotex users, 5. settlement of transactions and fees, 6. gateway to external data bases.

Since then, the development of the Swiss videotex system (see Figure 1) has become more and more decentralized: two new main computers were installed in 1985 in Zurich and in 1988 in Lausanne. From the beginning of 1989 onward, the PTT has used twelve additional telematic access-processors (TAPs) as frontend-processors or detached-concentrators, which function as gateways to external data bases. More than 80% of the videotex traffic now is settled by external data bases.

The main reason for this decentralized development is political. *SVIPA - the Swiss Viewdata Information Providers Association -* successfully tried to reduce the PTT's role to that of connecting the information-providers with the information-users by offering and operating the necessary telecommunication infrastructure. Therefore, the *Federal Videotex Decree* limits the role of the PTT as an independent information provider: Data base services are allowed only for public authorities and non-profit organizations.

AMOUNT AND TYPE OF SYSTEM USAGE

Quantitative aspects. Table 1 summarizes the main quantitative data regarding the development of Swiss videotex. During June 1990, 73,000 users were subscribed to the Swiss videotex, making 1.1 million calls to the system; that is 15 videotex sessions per subscriber per month. The average session length is 12 minutes. Videotex was used for 80 millions minutes in the first half of 1991, that means approx. 200 minutes by each videotex subscriber per month.

In 1985, 82% of the videotex subscribers were from the German speaking and 18% from the French speaking part of Switzerland; today, 70% are from the German, 27% from the French and 3% from the Italian language region (PTT survey, 11.9.89). This corresponds quite well with the distribution of the Swiss population. In 1991 about two thirds use videotex privately at home while 35% are business users. The dominance of business users was much higher during the test-phase in 1975, approx.70%.

The strong increase of videotex subscribers in the last three years is on the one hand the result of better and more interesting information possibilities provided by the Swiss videotex system, e.g. electronic home banking or PTT's *Telegiro* and electronic telephone directory. On the other hand, it seems to have been strongly stimulated by the PTT which has vigorously lowered videotex tariffs.

Another factor is that more and more subscribers use videotex as an integrated part installed in their personal computer: 43% of all videotex users in June 1991.

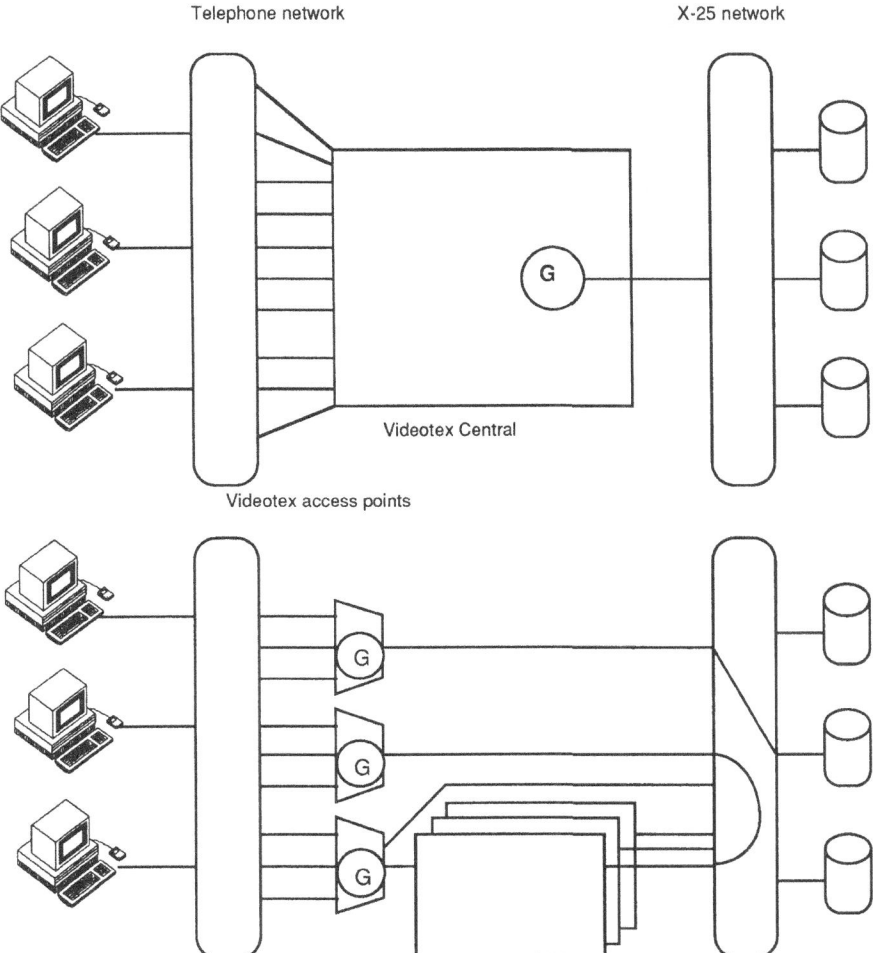

Figure 1: Development of technical structure of Swiss Videotex

Table 1. Subscribers, Information Providers, External Data Bases and SystemUse

	Number of Subscribers	Information Providers	External Data Bases	Use of System in minutes
Sept. 1984	296	90	13	-
end of 1984	970	127	19	-
end of 1985	2668	216	22	5,447,225
end of 1986	4187	275	32	9,855,874
end of 1987	7635	332	35	16,066,507
end of 1988	14474	332	41	37,638,223
end of 1989	35304	393	50	77,334,629
end of 1990	60315	523	61	134,425,090
July 1991	75205	570	61	93,873,067

Qualitative aspects. In 1989, six qualitative group interviews with a total of 60 videotex users and a telephone survey of 100 private and business users (IBFG 1989) each show various *social uses of videotex*: 1. the electronic PTT telephone directory is by far the most important information provided by videotex, followed by 2. home banking / telegiro, 3. up-to-date information (weather, finance information, leisure events etc.) and 4. communication (mail boxes, dialogue, clubs). According to the answers made in the qualitative interviews, tele-shopping still plays a minor role: there are not many interactive possibilities on the Swiss videotex system, ordering is expensive, complicated and takes longer in comparison to personally ordering by phone.

Despite this broad range of functions, most videotex users still are *very critical of the system*. Widely stated complaints are: speed, complicated and troublesome search process, and inadequate topical search procedures. In addition, much information and many (private) advertisements are not up-to-date and too many information-providers offer only one entrance page without further information.

Different *types of videotex users* seem to be developing. On the one hand there are purposeful and monofunctional users who use videotex only as electronic telephone directory or for home banking; on the other hand there are videotex subscribers who use the system in a more multifunctional and more experimental and explorative way. The usage of the medium videotex itself seems to have *intrinsic value* for the latter group. Whereas for the first type videotex has only an *extrinsic value*, namely as a means to get certain information quicker, cheaper or more up-to-date in comparison to other media.

Personal interviews with 150 videotex users (32% private, 47% business and 20% private and business) who *cancelled their videotex subscription* between June and September 1988 reveal the following arguments. Videotex is: 1. not adequately useful (47%), 2. too expensive (28%), 3. too slow (27%), 4. lacking offers (15%), 5. showing technical problems (15%), 6. raising difficulties with the electronic telephone directory (14%), 7. too complicated (12%). "*What would have to be improved to get you to subscribe again?*" : 1. Speed enhancement of videotex (45%), 2. improvement of the search process (32%), 3. more useful offers (30%), 4. lower hardware prices (25%), 5. more user-friendly operating routines (17%), 6. lower telephone taxes (12%), 7. more up-dated information and services (9%).

SERVICES OFFERED

In 1991, 570 information providers and 61 external videotex data bases has been connected to the system. Table 2 displays the main menu of the Swiss videotex system and the entries of the "topical menu". It also shows how many information providers are listed under each topical entry in 1991.

Table 2. Information Provided by Swiss Videotex System

Entrance Page of Swiss Videotes System			
1985:	**1991:**		
1# Table of Topics	1# Around Videotex		
2# Key Words	2# Table of Topics		
3# List of Subscribers	3# List of Subscribers		
4# List of Information Providers	4# List of Information Providers		
5# List of External Data Bases	5# Videotex International		
6# Vtx-Helps for Users	6# Electronic Phone Book		
7# Information about Vtx-Testphase	7# Futher Telecommunication Services		
8# Mail Box	8# Mail Box		
9# System Functions	9# User functions		
Table of Topics (*Entrance Page: 1#(1985 / 2#(1991))*)			
1985:	**1991:**		**Entries**
1# Construction, Real Estate	11# Labour, Profession	37	4
2# Living, Household	12# Education, Instruction, Science	36	4
3# Garden, Animals, Agriculture	13# Leisure, Sports, Entertainment	116	12
4# Food	14# Press, Books, Radio, TV, Cinema	69	7
5# Clothes, Jewellery, Watches	15# Couselling, Church	11	1
6# Electronics, Foto, Films	16# Garden, Animals, Agriculture	14	1
7# Medicine, Health, Cosmetics	17# Consumption, Delivery, Commerce	75	8
8# Banks, Insurances	18# Living, Construction, Real Estate	27	3
9# Trades, Industry	19# Encyclopaedia, Guides, Phonebooks	41	4
10# Delivety, Commerce	20# Arts, Culture	32	3
11# Services	21# Tourism, Hostellery, Weather	62	6
12# Computers, Office Electronics	22# Traffic, Transports	40	4
13# Trade Information	23# Craft, Trade, Industry	36	4
14# Labour, Profession	24# Public Offices, Politics, Taxes	12	1
15# Education, Instruction	25# Organizations, Clubs	37	4
16# Couselling, Social, Church	26# Medicine, Health	24	3
17# Press, Radio, TV	27# Banks, Finance-Info, Home Banking	61	6
18# Leisure, Sports, Entertainment	28# Insurances·	13	1
19# Arts, Culture	29# Services	127	13
20# Tourism, Hostellery	30# Dialog-Services	86	9
21# Traffic, Transports			
22# Weather			
23# Authorities, Politics, Justice		956	
24# Organizations, Associations	*Sum (multiple entries)*		100%

But one must know that many of the entries e.g. in *13# "Leisure, Sports, Entertainment"* refer to umbrella services, offered by publishing companies, videotex-agencies, banks etc. that provide among a number of other things some pages with games.

About a third of the 30 *external databases*, connected to videotex in 1987, have been installed in medium and large Swiss banks. Additionally, a fourth of the external computers are owned by computer companies (IBM, Siemens, NCR etc.) offering computing-, storage- and system-developing services to their clients or the EDP-departments of big publishing companies. The last 40% of the external computers connected to videotex are operated by specialized information- and service- providers, e.g. insurance companies. Worth mentioning is the external computer of the PTT that provides up-to-date telephone numbers and adresses of all telephone subscribers.

Table 3. Information Providers to Swiss Videotex System (1990/91)

	Information Providers:	absolute	percent
1.	Leisure, Tourism, Transportation	68	16
2.	Publishing Companies, Media	56	13
2.	Commerce, Delivery	58	13
4.	Mail Boxes, Communication, Clubs	47	11
5.	Banks, Finance-Information	45	10
6.	Electronics, Computers	34	8
6.	Videotex Agencies	33	8
8.	Associations, Clubs	14	3
8.	Authorities, Communities	12	3
8.	Insurances	12	3
11.	Agriculture	11	2
11.	Production, Industri, Trade	9	2
11.	Advertising, PR, Marketing	8	2
11.	Health, Medicine	7	2
15.	Others	17	4
	Total	431	100 %

Leisure, Tourism, Transportation: Airlines and flight information of Zurich Airport, tourist information about different regions and resorts, travel agencies, several hotels, guides of the larger cities in Switzerland (Berne, Zurich, St.Gallen, Basel), weather forecast, several dialoge-clubs.

Publishing Companies and Media: More than a dozen of the medium and larger newspapers offer not necessarily up-dated headlines. The bigger publishing companies have their own umbrella information services, e.g. with city-guides. The biggest external data base is *"Telepress"*, the videotex agency of a group of big newspapers (Tages-Anzeiger, Bund, 24heures, Basler Zeitung, St.Galler Tagblatt, Orsysta).

Commerce, Mail Ordering: Several of the bigger department stores offer possibilities for tele-shopping; some publishing companies allow to order books; electronic auto markets are offered. But in general, tele-shopping is still underdeveloped and the mail-order business offers only a small proportion of the available products on videotex. A few of the Swiss settlements of big international motor campanies (e.g. Volvo, Mercedes, Nissan) have elaborated, closed videotex applications. These allow the ordering of spare parts by licenced garages.

Mail Boxes and Communication: In the last few years, many new videotex agencies offer mail boxes and communication services. Especially prominent are more than a dozen mostly erotic communication clubs that have provided the Swiss videotex system wide media publicity lately. Insiders estimate that up to 70 % of the total videotex system time today has to do with messaging and communication.

Banks and Finance Information: Almost all of the bigger national and regional Swiss banks are information providers. They offer at least information about services, interest rates, stock exchange quotes, investment advices etc. Many of them (more than 30 in May 1991) have interactive programmes that allow electronic home banking. More than two thirds of the videotex users are subscribers to the home banking services. But most popular is *'Telegiro'*, PTT's electronic banking service with more than 15,000 subscribers. It is said that Swiss banks invested the biggest portion of all in videotex and are the strongest supporters of the system.

Electronics and Computers: Big computer companies like IBM, Alcatel STR, NCR, Nixdorf, Radio Schweiz offer external data base services (hard- and software) for videotex information providers that have no storage facilities.

Authorities and Communities: As a result of the *"Project of Communication in Model Communities"*, initiated and sponsored by the PTT, several small and medium communities have developed *"community information systems"*: Frauenfeld, Lausanne, Locarno, Maur, Nyon, Sursee. Beside more commercial information, videotex users can access the results of the last community vote as well as other actual community information e.g. planned construction projects, deaths or births, garbage disposal sites etc. In the case of Frauenfeld, an evaluative study (IBFG 1991) showed that the 22 pages which were offered exclusively by the community council, ranked fourteenth (1%) of all videotex pages accessible in the community information system, but positioned fifth (6.3%), in total pages accessed. - Nevertheless, the most used service is still PTT's electronic telephone directory with about 600,000 monthly calls.

Insurances: The bigger Swiss insurance companies offer general information about insurance problemes and special information about their services.

Agriculture: Several agricultural associations and agricultural advice centers offer up-to-date information for farmers.

Production, Industry, Trades: Up until now there are not many companies that use videotex as a medium for PR. Exceptions are some big multinational companies like Nestlé, Ciba-Geigy and Hoffmann - LaRoche.

Health and Medicine: There are only a few health data bases (e.g. AIDS statistics) and general health information services.

In summary: There have been several new offerings in the last few years that have significantly enhanced the popularity and usefulness of the Swiss videotex system: the up-to-date PTT telephone directory has been accessible by videotex since October 1986 and the PTT introduced its electronic *Telegiro* service in October 1988 that is similar to the private home banking services. The timetable of the Swiss railways and the postal bus service'has been accessible since June 1991. But in general, information and services provided by the Swiss videotex system has not been overwhelming and is still not convincing. The range of applications is not wide enough; the individual services are often not comprehensive or up-to-date; and user-helps or visual design of many videotex programs are in need of improvement.

COST STRUCTURE

Since the introduction of videotex in Switzerland the question of tariffs is still a source of conflict namely between the PTT and the SVIPA, the Swiss Viewdata Information Providers Association. Videotex subscribers have to pay for several things: a) fixed monthly telephone connection fee, b) fixed monthly videotex connection / modem fee, c) fee for renting a videotex terminal, d) actual use of the telephone line by amount of time, e) variable costs for tax-required pages. All these fees are collected by the PTT.

Once-off installation fee: A new videotex subscriber has to pay the PTT SFr. 60.-- (34.25 ECU) for the link up to the videotex system; this does not include the potential costs for the actual installation by a local service company.

Basic telephone connection fee: Videotex in private households is usually connected to the rented telephone line. The new dedicated videotex terminals also include a phone. In Switzerland, every owner of a telephone has to pay a fixed monthly fee of SFr. 20.-- (11.5 ECU).

Fixed monthly videotex connection fee: It the beginning videotex was operated by a decoder connected to the television set and a modem connected to the telephone. The videotex user had to pay a fixed monthly videotex connection fee including the modem-fee of SFr. 12.-- (6.8 ECU). Since most of the videotex subscribers now use either a videotex terminal with integrated modem or a videotex card in their personal computer, the PTT has removed this videotex connection fee at the beginning of 1989.

Tax for renting a videotex terminal: Few videotex subscribers purchased terminals at quite high prizes. Because the PTT nowadays offers cheap terminals to popularize videotex, most private videotex users rent a terminal. The cheapest terminals for rent, similar to the French Minitel, are between 9 and 25 SFr. (5 - 15 ECU) per month.

Connect time to the videotex system: The videotex user has to pay a distance-insensitive tariff of SFr. 7.50 (4.25 ECU) per hour permitting access to the videotex system regardless of the physical location of the videotex user in Switzerland. To further stimulate videotex diffusion the PTT lowered this fee to SFr. 3.-- (1.7 ECU) at the beginning of 1988. This tariff corresponds to twice the local-zone telephone tariff.

Tax-required pages: Information providers can charge videotex users by time, which is similar to the french kiosk-system, or by page. Both tariffs are variable and are depending on the policy of the information providers. The highest price they are allowed to charge is SFr. 9.95 per page (5.7 ECU). Information providers have to display on a non-taxed page the cost for the tax-required pages in advance. These costs are charged on the regular phone bill. In 1989 the private information providers earned SFr. 2 million (1.2 million ECU). This amount increased to SFr. 10.8 million (6.2 million ECU) in 1990. In average, each subscriber spent per year approx. SFr. 250 (143 ECU) for those services.

Costs for information providers: The costs for information providers break down into several different elements. Operators of external data bases have to pay the PTT the costs for transmitting the information on the telepac network (plus several additional tax components). For the moment these are restricted to an upper limit of SFr. 2500 (1430 ECU) per month, which benefits the big information providers. The costs of an information provider to develop, store and periodically update a videotex service on an external data bank are certainly not included.

Today, Swiss videotex is still operating at a loss. Until now, the PTT alone invested about SFr. 70 million (40 million ECU) and earned SFr. 8 million (4.6 million ECU) in 1990 (mostly telephone taxes), meaning, only about 35% of the effective costs are covered by the subscribers and providers of the information'. As a temporary arrangement based on a the Federal Accomplishment Duty, it has been proposed that the PTT will be allowed to subsidize videotex further. When the new Federal Telecommunication Law will be enacted, probably in 1992, the PTT will have to replace the videotex decree by a new one. Since cross-subsidizing between different services will no longer be allowed, videotex tariffs will rise significantly.

History of Videotex in Switzerland

Like in other European countries the introduction of telephone-based videotex as an enhancement of the telecommunication capabilities has proved to be more difficult than originally expected. Three main phases can be distinguished in this *process of media innovation*: a first small-scale phase of technical experimentation, a second phase of testing the market potential and evaluating the potential negative impacts of this new medium on a broader basis, and a longer third phase after the official introduction of the Swiss public videotex system: the early exaggerated hopes were slowly replaced by pragmatic expectations.

TECHNICAL EXPERIMENTATION

At the *end of* 1979 the Swiss postal services started a closed, four-year small scale *technical pilot project*, based on the British Prestel system. 130 information providers stored 6000 videotex-- pages in a main PTT computer at a cost of about SFr. 1.5 million (850,000 ECU) . As a result and without a significant public debate, the PTT applied for a public field-trial and the Federal Government gave permission on the 3 of February, 1982.

MARKET EXPLORATION PRIOR TO SOCIAL EVALUATION

1. September 1983 was the official beginning of a two year long *field-trial* based on European CEPT standard. Due to a shortage of available vtx-decoders the testing phase started not until summer 1984. The PTT decided against the project offered by IBM which was similar to the German *Bildschirmtext*, and chose the more decentralized system of *Standard Telephone & Radio (STR)*. This system is based on external data bases and the main functions of the two PTT computers in Berne and Zurich consist mainly in storing the basic protocols and providing access and gateway to the external data bases.

Only 220 organizations delivering information and 2,700 subscribers to the videotex service, 700-800 of them in private households and 1500 in offices, were connected to this computerized public data source by the end of 1985. Until then the PTT invested about SFr. 50 million (28.5 million ECU) and the information providers another 40-50 million in the Swiss videotex system (other sources quote a total investment of SFr. 150 million (85.7 million ECU). Based on a concept with concentrated test-markets, the main goal for the PTT was to explore the needs of information users and the market possibilities of videotex in Switzerland (Itin 1983).

Table 4. Development of Videotex in Switzerland

Dec. 1977	- First presentation of Vtx by STR at PTT board of management
March 1979	- Formation of SVIPA (*Swiss Viewdata Information Providers Asc.*)
from 1980	- Loosening of PTT monopoly on modems and cheaper tariffs for modem-leasing and for videotex subscription per month
Nov. 1979 - Sept. 1984	- Four-year closed technical pilot-test based on British Prestel with 250 subscribers, 61 firms and a PTT main computer with 8000 pages
3. Feb 1982	- Permission of Federal Council to public pilot-test with evaluation
1. Sept. 1983	- Official start of the two-year operating testphase based on CEPT
Sept. 84 - end of 1985	- Delayed test-phase with PTT main computers in Zurich + Berne; 1500 vtx subscribers and 216 info-providers
	- Reduced vtx subscription fee of SFr. 12 / 6.8 ECU (modem tax included) and reduced connect tariffs: SFr. 7.70 / 4.25 ECU per hour
	- Result: "*high expectations at the start, disenchanted experiences during operation test, pragmatical perspectives for the future*"
until 1985	- Total investment of SFr. 150 mio. (85.7 Mio. ECU) for Swiss vtx
end of 1985	- Public notification to proposal of videotex decree
Oct. 1986	- PTT telephone directory available on videotex
1. Jan. 1987	- Official start of Swiss videotex, based on Government Decree, with 4187 subscribers and 275 information providers
during 1987	- PTT advertising campaign promoting videotex
Feb/Mar 1987	- Installation of front-end processors to relieve the main vtx-computers by gateway functions
1.Jan. 1988	- Reduced taxes: from SFr. 7.50 to 3.00 per hour (4.25 - 1.7 ECU)
Feb. 1988	- Futher PTT main-computer in Lausanne
June 1988	- Vittel 100: first "swiss made" vtx-terminal by ASCOM
Oct/Nov. 1988	- Test phase with 110 videotex terminals installed in public places
3. Oct. 1988	- Introduction of "*Telegiro*" by the PTT: vtx for postal accounts
1988 - 1992	- "*Communication-Model-Communities*" by PTT to test needs for new services in 12 communities with several videotex applications
1. Jan. 1989	- Monthly connection fee abolished; cheapest vtx-terminal rented by PTT for 9 SFr. (5 ECU) per month
	- Press reports on flourishing sex-dialogues on videotex
end of 1989	- Videotex has 35000 subscribers; information providers earned SFr. 2 Mio. (1.15 Mio. ECU) with tax-required pages
1990	- Futher PTT main-computers in Basel and St. Gallen, possibilies to switch to foreign vtx services in Germany, Austria and Luxembourg
	- National promotion campaign by PTT for videotex
end of 1990	- Videotex has 60000 subscribers; the PTT forecasted 1983 10 % of the swiss households (= 260000) in 1990
	- Private information providers earned SFr. 10.8 Mio. (6.2 Mio. ECU)
June 1991	- Timetable of Swiss railways on videotex

An *evaluative study*, required by the federal government and paid by the PTT (SFr. 200,000 / 115,000 ECU), analyzed the market possibilities, diffusion and acceptance of this service by the public, the potential beneficial and/or harmful effects on the other media, especially the press, on the economy and on private households (Meier & Bonfadelli, 1987).

Results: The early videotex users are innovative adopters, 95% are men, 73% between 30 and 49 years old and 63% have college and university education. Business and not private usage of videotex dominate. Videotex is mostly used for telebanking .(45%). Other uses include tele-shopping (9%), entertainment (7%), and travel/tourist-information (5%). Only 40% of the users are satified with this electronic information service. The main reasons for rejection at the moment are: *technical deficiency* (search process too complicated, delay of response), *lack of useful content and expensive* (lack of cheap videotex decoders). - The initial euphoria at the introduction of videotex in Switzerland has vanished and a pragmatic and sceptical view of future development dominate today (IBFG 1985).

In November 1985 the Federal Government announced the proposal of a *videotex decree* and carried out a public notification. The proposed decree defined videotex as an open public service with the main computer and the telephone network provided by the PTT. It allowed only modems and terminals that were permitted by the PTT, required that tax-obligatory pages were announced in advance, and regulated data and user protection.

The results of the notification have been very controversial. The SVIPA and the liberal parties criticized that the decree also enclosed *private* videotex services and they opposed any regulations to protect personal data. The left-wing parties and other organizations asked for a more extensive regulation on the level of a Federal Videotex Law with special emphasis on personal data protection.

As a consequence, the Swiss Government restricted the Videotex Decree to the public videotex service offered by the PTT; explicitly reserving for the future the possibility of regulations for private videotex services. The PTT is allowed to offer its own data-banks only to non-commercial organizations and public authorities. The PTT bills for tax-obligatory pages and then pays the private information providers. Furthermore, Art. 6-8 in the proposal, concerning matters of data protection were removed. Still remaining are several *content-orientated regulations*: forbidden are services and information that are a) unlawful, b) offend against moral senses and public authority, and c) contain hidden advertisements.

FROM EXAGGERATED HOPES TO PRAGMATIC AND MODEST EXPECTATIONS

The Videotex Decree was passed by the Government on *January 1, 1987* and the videotex service went in full operation with a mere 4200 subscribers. As a consequence, the PTT started an *advertising campaign* in 1987 (more than 100,000 SFr., 57,000 ECU) to increase public awareness and interest for this new electronic information service: *"Videotex the ingenious phone"*. To enhance the popularity of videotex the PTT lowered, at the end of 1987, the connect-time tax from SFr. 7.50 to 3.00 (4.25 to 1.7 ECU) and the monthly subscription from SFr. 50 to 38 (28.5 to 21.7 ECU) and subsequently cancelled it. Cheap videotex-terminals could be rented for SFr. 22 (12.5 ECU) per month and new services like the electronic telephone directory and *Telegiro* (electronic charging of PTT accounts) have been introduced. In a further step, the PTT initiated and sponsored (at a total cost of SFr. 257 million, 147 million ECU, SFr. 144 million, 82 million ECU payed by the PTT) the so called *"Communication in Model--Communities Project"*: new telematic and telecommunication possibilities are actually tested in twelve communities. Especially new videotex applications like community information systems were to be developed. - As a result, the subscribers to the Swiss videotex doubled from 7,600 at the end of 1987 to 14,000 at the end of 1988 and reached 73,000 in June 1991.

Main Actors and Interests Involved

The main actors and interest groups involved in the development of the videotex system are visually displayed in the so called "*PTT-videotex-flower*" (see figure 2) that was included in the initial conceptual papers of the PTT: 1. Videotex terminal dealers, 2. producers of terminals and decoders for videotex, 3. electricians for the installation of videotex, 4. politicians and authorities, 5. hardware suppliers of the videotex system, 6. videotex software developers, 7. information providers, 8. videotex subscribers, 9. PTT. Only a few of these nine actors or interest groups however had any real influence on the development of the Swiss videotex system.

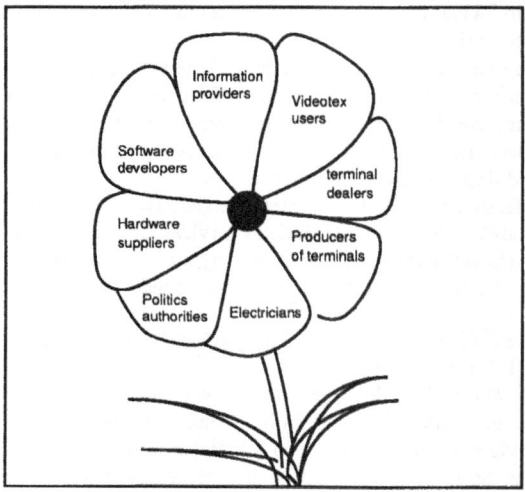

Figure 2: PTT-Videotex Flower: Main actors involved in Swiss Videotex

Swiss Postal Services (PTT): Videotex was introduced as a reaction to pressure by the telecommunication industry and as a way to increase traffic on its telecommunication system. In view of the planned, new integrated digital services network ISDN evaluation and stimulation of user interest was another aspect. The PTT realized from the beginning, that its role would be restricted to the function of providing and operating the telecommunication system as an infrastucture for videotex. Therefore, the private industry and its various pressure groups would not allow the PTT to act as an information provider. This was the main reason for choosing a decentralized system instead of the German or French videotex system. A further question had been the market of the so-called end- or user-devices. The PTT traditionally maintained a monopoly. Today videotex-users are allowed to buy their own modems and connect to the videotex system by videotex-decoders that are built into their personal computers. The PTT still lets videotex terminals, but in the future, videotex will become an integrated part of the personal computers.

International manufacturers of computers and telecommunication systems: They had been interested in new innovative markets as a result of the stagnation of the telephone sector. *Standard Telephone & Radio (STR)* won against the centralized system that IBM offered and delivered the videotex main computers to the PTT. Other computer companies interested in videotex (like NCR and IBM) now offer integrated system solutions and are especially strong in the home-banking sector. Swiss manufacturers of telecommunication devices, traditionally strong as clients of the PTT, tried to construct their own videotex terminals, but failed. Therefore, the PTT buys all videotex terminals in Germany.

. *Swiss Viewdata Information Providers Association (SVIPA)*: Founded in March 1979 as an industrial interest group with the explicit goal of promotion, development and representation of the interests of the information providers to the Swiss videotex system. SVIPA was and still is the most important partner of the Swiss PTT. Its six permanent project teams on topics like marketing, tariffs, security, law, system technics, and search structures worked out numerous proposals that were often accepted by the PTT. There still is an ongoing, close cooperation between the PTT and the different project teams of the SVIPA in the development of the Swiss videotex system. From the beginning, SVIPA tried to limit the influence of the PTT and successfully demanded that the PTT act only as provider of the telephone network and is not allowed to offer own data-base services. Other important demands have been: exclusion of the new medium videotex from the proposed Radio and Television Law in 1983; successful limitation of the Federal Videotex Decree to the public videotex system only and deletion of the proposed data-protection measures in 1986; abolishment of PTT monopoly on end-devices; several finally successful attempts to reduce videotex and telepac tariffs for information providers and videotex subscribers.

Swiss newspaper publishers: Interests of newspaper companies in videotex were more defensive than offensive. The main reason was a strong fear of advertising losses. Therefore this was one of the explicitly stated research goals of the videotex evaluation study. As a consequence, a group of important newspaper publishers started to operate the external videotex data base *"Telepress"* and other newspaper publishers offer their own information services. In most cases these services combine daily headlines with electronically advertised markets. But until now, no publisher tried to offer an integrated package, consisting of advertisement in both newspaper and on videotex.

Federal Government and local authorities: The Federal Government did not play a very active role, either in the development or in the regulation of the public videotex system in Switzerland. Its role can best be described as balancing the interests of the different actors interested in videotex. One factor was certainly the *unsatisfactory and unclear legal base* for the introduction of videotex. Nevertheless it was certain that the *Federal Council* owned the prerogative for postal & telecommunication services (Art. 36 BV) and that the PTT were the solicitor of this prerogative; unclear however was the authority of the PTT concerning experimentation with and introduction of new telecommunication services, for they are not mentioned in the old Telecommunication Law dating from 1922.

Instead of discussing these basic questions, the *Federal Council* acted very pragmatically: as a consequence of the initial fears of most Swiss publishing companies the PTT was forced to carry out an evaluative study that included not only marketing questions, but also questions of technological assessement too. But even prior to the start of the test-phase, the general director of the PTT, Werner Trachsel, proclaimed that the main goal of the evaluation was not to decide for or against videotex but how to introduce it. In a next step the Federal Government formulated the first proposal of the Federal Videotex Decree, but removed all data protection measures due to the criticism from industry and liberal interest groups. The Federal Government officially allowed the introduction of videotex at the beginning of 1987. On the local level only, several communities took a more active part in the PTT sponsored project of *"Communication in Model Communities"* and developed community information systems on a very limited scale.

Political parties and public interest groups: The main criticism formulated by the Social Democratic Party (SP), Trade Unions (ARBUS, VPOD), the Left-wing Journalist Associations (SJU, SSM) and the Media Councils of the protestant and catholic churches, was the absence of any public discussion about the possible, negative social impact of videotex, its introduction and regulation. In fall of 1983, these groups presented a first critical comment on videotex at the annual videotex conference in Basel. Later, in spring 1984, the Media Institute of the catholic church organized a public conference on the topic of videotex and in 1985 a reader with the titel *"Cold Communication: The Million Game Around Videotex"* (Frischknecht 1985) was published as a critical contribution to the Federal notification on the proposed videotex decree. The main arguments against videotex were: videotex as a rationalizing means (e.g. for the banks and the industry) will destroy many jobs; it will enhance control over consumers by delivering detailed consumer data; it will lower interpersonal communication, foster personal isolation and further strengthen the concept of a more and more rationalized construction of the everyday world. Based on these arguments the following claims have been proposed: introduction of videotex has to be based on an evaluative study and has to be discussed critically in public; no public subsidies for the forced introduction of videotex; guaranteed options between different (electronic and non electronic) public services and no forced choices e.g. by differently prized services; introduction of videotex has to be based on Federal Law and this Videotex Law has to guarantee a minimum of user- and data-protection and clear regulations concerning the separation between commercial and non-commercial information; non-profit information providers have to be subsidized.

To sum up: The PTT has been the most active and decisive actor in the process of development and installation of the pulic videotex system in Switzerland. But it was able to maintain this position only by acting in small pragmatic steps that harmonized with the interests and options of the SVIPA, the only organized interest group of the information providers (above all the big Swiss banks) and the operators of external data-bases. During this process of media innovation the PTT could play a relatively active part only by accepting its role as network operater, the strict division between transport and content of the information and the concept of videotex as a decentralized system, based mainly on external data bases, with a uniform page-based, format-- standard for the information user. The two main reasons for the SVIPA to accept the central role of the PTT were on the one hand, only the PTT could offer cheap distance independent telephone tariffs, and on the other hand, the possibility that the PTT as a centralized organization could charge the information users vicariously on behalf of the information providers.

Conclusions: Problems, Conflicts and Solutions

Technology: While the pilot-test with videotex started with a central computerized data base kept by the PTT, the technical development and even more important the mediapolitical context enforced a decentralized concept of the Swiss videotex system. PTT's videotex computers would only handle identification of the videotex subscribers, store the protocols and function as a gateway to provide access to external commercial videotex data bases. Today more than 80% of the videotex connecting hours are settled by external commercial data bases. From a user point of view, videotex technology started as an additional enhancement of the television set in the home entertainment context, but rapidly developed from a small videotex-terminal to an independent office or home telecommunications device. In a third phase, there is a tendency in the home computer that already includes display, keyboard and more important electronic storage and intelligent search possibilities. As a consequence of this development the new marketing strategy of the PTT is targeted especially to owners of personal computers.

Medium: Videotex was propagated in its early stage as a mass medium primarily with information functions for the private user. Instead of this concept, videotex was first adopted mainly in a (semi-)professional context: the testing phase showed that 66% of the videotex subscribers used videotex for commercial, 27% for private use and 7% for both. It took quite a long time, and some new applications aswell, for the broader public to discover videotex as a medium for (interpersonal) communication and as a new electronic tool to accomplish daily tasks like banking or searching for an unknown telephone number. As a consequence the portion of private videotex users rose to 65% between 1985 to 1991 and the share of mainly commercial users declined from 66% to 35%. Another fact was that the multilingual Swiss videotex had a strong competition in the French part of Switzerland from the French Minitel. But from September 1985 to September 1989, the proportion of users from the French speaking part of Switzerland rose from 18% to 27% and the portion in the German part of Switzerland decreased from 82% to 70%. This is partly due to the possibility of switching easily between languages in the Swiss videotex system: almost all information providers present their applications in three different languages.

Contents: Information, communications and services accessible by Swiss videotex are very diversificated and their use varies according to the very different needs of the various user groups. But there have been some remarkable changes: Information providers overestimated the public's need for large amounts of journalistic day-to-day and other more specialized background information. Contrary to these expectations, the testing phase proved that private videotex users did not use videotex as a fast mass medium and did not want actively to search and to pay for more specialized background information. Whereas other services proved to be more useful to meet existing needs and satisfying them at lower cost. This is especially true for the electronic telephone directory and home banking or PTT's telegiro services. These contents have pushed considerably the diffusion of videotex in the last years.

There still remain significant problems: the more or less arbitrary division between the PTT and the SVIPA concerning the two functions of information provision and information transmission has defined videotex as an open and practically non regulated and non transparent market place. This situation can be very troublesome for the videotex user because of the almost worthless topical search procedures. The main reason lies in the fact that most of the information providers are listed with several topical entries and the videotex user is never sure what he will find at the end of his/her multistep search process. There is no supplier who feels responsible for the public needs and who also has the power to sanction some of the existing misuses by the commercial information providers.

Pricing: To increase popularity of videotex, the Swiss PTT had to lower the videotex fees several times. Although the SVIPA wanted a local rate, the PTT applied a fix regional tarif (20 km), which was reduced on 1 January 1988 from 7.5 SFr. to 3 SFr. (4.25 to 1.7 ECU) for one hour. A next step was to cancel the initially required monthly connection fee. This user friendly prizing system was enforced by a marketing strategy based on an ongoing information campaign, installation of public videotex terminals, and the "model communities" project.

Summary: As in other European countries, diffusion of videotex in Switzerland was clearly slower than expected and it is not yet clear whether a substantial segment of the population has or will develop information needs active and strong enough to stimulate subscription to and persistent usage of this new electronic interactive medium. Nevertheless, particular to the Swiss situation seems to be that videotex eventually succeeded at a modest level, though tiny pragmatic steps. Relevant factors are on the one hand low terminal prizes and cheap connect time fees, made possible by crosssubsidisation, and on the other hand the close and continuous cooperation of the PTT with the SVIPA and other important videotex information-providers, especially the Swiss banks.

References

Bonfadelli,H. & Hättenschwiler,W. (1989). 'Switzerland: A Multilingual Culture Tries to Keep its Identity', in L. Becker and K. Schönbach (eds.), **Audience Responses to Media Diversification**. Hillsdale, N.J., Erlbaum Publ, pp. 133-157.

Frischknecht, Jürg(ed.) (1985). **Kalte Kommunikation. Der Millionen-Poker um Videotex und andere Neue Medien**. Basel, Lenos Verlag.

Interdisziplinäre Berater- und Forschergruppe Basel (1989). **Vtx-Labor: Auswertung der 6 Gruppengespräche**. Basel, IBFG.

Interdisziplinäre Berater- und Forschergruppe Basel (1991). **Angebotsanalyse 'KMG Frauenfeld'**. Basel, IBFG.

Itin, P. (1983). 'Die Strategie zur Einführung von Bildschirmtext in der Schweiz'. **Media Perspektiven**, 2, pp. 86-93.

Knecht, Wilhelm (1984). **Videotex das neue Medium**. Zürich, IBO-Verlag.

Meier, W.A. and Bonfadelli,H. (1987). 'Comparative Analysis of Videotex Pilot Project Studies in Three European Countries'. **Telematics and Informatics**, 4, pp. 75-83.

Orell Füssli AG (ed.) (1990). **Schweizer Videotex-Guide**. Zürich, Orell Füssli.

SRG Forschungsdienst (1991). **Jahresbericht des Forschungsdienstes 1990**. Bern, SRG.

CHAPTER 7

BELGIUM: BETWEEN MONOPOLY AND COMPETITION

François Pichault & Marc Minon
LENTIC - University of Liège

Introduction

In Belgium, more than anywhere else, prospects in the field of videotex are especially favorable, given the existence of the Teletel Programme in France, a neighboring country whose language is spoken by a sizeable minority of the Belgian population. This vicinity leads to a number of different competitive advantages.

- Generally speaking, the successes and failures of the Teletel Programme pave the way for the definition of a coherent strategy (massive distribution of paying terminals, monthly renting at reduced rates, a simple norm, integrated terminals, telephone kiosque, electronic mail, etc.) and make it possible to reduce uncertainty for any national operator.

- Moreover, the existing 6,000 000 or so terminals do permit economies of scale: first on the terminal which can be made available, in the Teletel/ASCII version, at conditions close to the present marginal cost for France Telecom — less than BEF 7,000 (ECU 166) — but also on the network equipments and host computers (hard and soft).

- Neither should one underestimate the potential importance of "advertising effects" on the Teletel Programme, owing to the presence of the French networks on the Belgian cable, the French magazines and daily papers read by French-speaking Belgians, tourism, etc.

- Finally — and most important perhaps — the existence of a considerable number of terminals in France could ensure a preliminary market for Belgian services with international outlooks, and the existence of more than 10,000 services in France could increase the potential interest of videotex for early Belgian users.

There is no denying that, under those conditions, the "chicken and egg" dilemma (Minon & Pichault, 1989) — to which the development of videotex was confronted in every country — could be confronted in a totally different way. And for the French-speaking part of the country at least (Wallonia and Brussels) the problem of the "critical mass" could be approached quite differently as well. However, as we will see, the corresponding reality does not really benefit from these positive elements.

The videotex landscape: a disappointing reality

A. USERS, TRAFFIC AND SERVICE PROVIDERS

On 31 July 1991, there were about 8,500 videotex subscribers in Belgium, mainly coming from the professional area and interested in specific applications relevant to their individual businesses (tourism, finance, news, etc.). This professional orientation corresponds to a deliberate decision made by RTT, the Belgian TT operator (see below).

H. Bouwman and M. Christoffersen (eds.), Relaunching Videotex, 85–98.
© 1992 *Kluwer Academic Publishers.*

Table 1. Evolution of the videotex subscribers and connection times in Belgium

Month	Number of subscribers	Connection time (in minutes)	Monthly connection time per subscriber (in minutes)
1988-01	2537	843427	332
1988-03	2867	963101	335
1988-06	3303	906714	274
1988-09	3795	940101	248
1988-12	4303	869344	202
1989-03	5234	1224936	234
1989-06	5864	1179433	201
1989-09	6096	1132878	186
1989-12	6365	1049440	165
1990-03	6520	1312164	201
1990-06	6806	1280500	188
1990-09	7082	1232770	174
1990-12	7742	1230840	159
1991-03	8074	1365439	169
1991-07	8518	1259484	148

(Source : RTT, Videotex Department)

Different lessons can be drawn from the table 1 concerning the users and their practices:

•   The total number of subscribers is increasing, but the rate of increase, whatever RTT's efforts have been, is now falling: + 70% in 1988, + 47 % in 1989, + 22 % in 1990. Moreover, the number of terminals, even in relation to the number of main lines, is much smaller in Belgium than in most of the European countries at a similar stage of economic development.

•   It is the evolution of traffic on the network — probably the best indicator of market evolution — that reveals the most disquieting results: between January 1988 and July 1991, the traffic increased by less than one third, significantly less even than the increase in the number of subscribers. This represents a dramatic fall in the average consumption per subscriber, which indeed went down from 332 minutes per month in early 1988 to 148 minutes per month in mid-1991.

The average consultation time per service has fallen as well and is now about a hundred hours per month. At such a level, it is obvious that, barring exceptional cases, the profitability threshold cannot be reached for services deriving their revenue mainly from traffic. On aggregate, revenue accruing to all Belgian service providers (not considering possible subscription to services) is estimated at less than BEF 2,500 000 per month (ECU 59,500), which clearly shows that the traffic generated on the network remains far below the level it should reach to guarantee decent conditions of profitability to the service providers.

There are presently 192 services available on the network, divided into three main categories: open (accessible to any videotex subscriber), restricted access and reserved services (limited to a set group of users). One can mention, in decreasing order, 20 services in financial sector[44] , 13 in press, 11 in retailing[45] and public sector, 10 in tourism, etc. But we must admit that the figure of 192 services does not reflect the situation accurately: it includes "transplanted" foreign services and the different levels of Minitelnet, giving access to French services for subscribers equipped with a Teletel terminal, the Belgian VAPs being multistandard. Moreover, services provided in the two national languages can be counted twice. Therefore, one can say that, four years after the beginning of the Belgian videotex experience, the supply of national services remains too low.

This rather disappointing situation results, in our view, from a specific configuration of factors that will be analyzed further (see '*Videotex: a mirror of RTT's policy in the NIT sector*', later in this chapter).

B. NETWORK CONFIGURATION AND TARIFF STRUCTURES

The Belgian network uses the Prestel norm but can support other norms (BTX or Teletel): it is one of its main competitive advantages. Its architecture is halfway between the centralized approach that characterized the early German, British or Dutch videotex projects and the French decentralized approach. Service providers may choose between different solutions: either they develop their applications on their own host computer or on the public one, with a cost depending on the memory size. The videotex network is organized around Videotex Access Points (VAP), a network management centre and the packet-switching network (called DCS). It can support about 15,000 users.

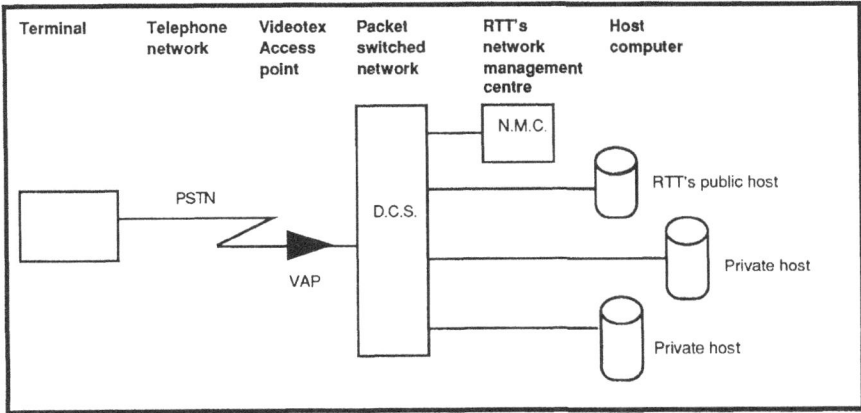

Figure 1: the RTT videotex network

Access to the RTT videotex network is dependent upon the payment of a subscription: BEF 300 per month (ECU 7.1) inclusive of taxes. Beyond that, the price depends on the complex interaction of several variables: volume of information transmitted, length of calls, rate for the VAP connection, etc. All this amounts, in actual fact, to an average price of about BEF 3.50 per minute (ECU 0.08). On top of this, which only represents the payment due to RTT as carrier, the user will also have to pay duties that private providers may levy for access to their services. As there are no rating levels, these duties, which RTT may include in the videotex charges, can vary considerably, although in principle not more than BEF 100 per minute (ECU 2.4).

More over, the user has to purchase his terminal (modem-card or integrated terminal) on the market, i.e. from retailers or suppliers. An integrated terminal (Prestel/Teletel/ASCII) sells at about BEF 18,000 (ECU 430) and can be rented for BEF 600 per month (ECU 14.3) inclusive of taxes.

A development in different "waves"

The Belgian approach has been fundamentally different from the French one, so that no advantage has been derived from the existence of the neighboring Teletel Programme. The latter was an early-starter, characterized by a voluntaristic attitude. The Belgian approach, on the contrary, has been slow and cautious.

A. FIRST WAVE: 1986-1989

Slow indeed: it was only in March 1986, several years after similar projects had been launched in the other countries, that the Belgian operating company started its own videotex system. Cautious also: having observed what was happening abroad, RTT definitively abandoned the "general public" orientation[46] and exclusively aimed at professional users, expecting, however, a significant progression in terms of subscribers. RTT chose the UK Prestel norm, the licence for which had just been obtained by Bell-Telephone, its main supplier. The initial investments were extremely limited (BEF 180,000 000 or ECU 4,285 000), consisting of the central unit of the system, the videotex software and the RTT data base.

In this context, the RTT videotex system condemned itself to relative confidentiality and it is not

---

[46] At the beginning of the 80's, RTT planned to undertake a videotex experiment for a period of 12 to 18 months, very similar to the French DGT experiment in Vélizy. Its goal was to investigate the technical feasibility of the system, the potential demand for it, and the legal problems surrounding the development of such a new technology. At that time, the data base was expected to contain 100,000 pages and to offer a similar kind of information to that available in other European experiments. For this purpose, the Belga press agency, the Belgian Railways, some banks, press groups, tour operators, etc. were contacted. As a first step, 400 households (200 in each linguistic community) were expected to participate in a free experiment. A further 200 professional users had to be supplied with the necessary terminals.

But given the poor results of similar experiments in other countries (Netherlands, UK, Germany, etc.), the initial project had been postponed year by year until its official launching in 1986. This reorientation towards professional targets clearly emerges from an interview with RTT's chief executive in 1985: "We have always preferred putting the brake on rather than speeding up", he said, "Foreign experiments, for example in the Netherlands, have warned us to be careful: an eventual demand does not always correspond to one's initial hopes. However, all is not negative: well-equipped access terminals may create, perhaps subliminally, some new needs. We shall draw our conclusions from the foreign experiments and begin on a small scale with about 600 access points. But they will be for professional use, not for households".

surprising that several competing initiatives were developed on the switched telephone network, on the packet-switching network, or on hired lines. One telling example is Bistel, the information videotex service of the Government — another competing public initiative, based on the Prestel norm — which appeared to generate a traffic equivalent to one third of the calls registered on the RTT network (see below).

This situation may be considered as the "first wave" of the videotex history in Belgium. It corresponds to a general orientation of the public videotex projects observed in most European countries at the same period towards potentially profitable targets (Pichault, 1987).

The BISTEL system

Parallel to the public RTT system, BISTEL (Belgian Information System by Telephone) is the videotex system of the Belgian Government. Resulting from the initiative of the Prime Minister W. Martens, it opens a videotex access to several public or private databases through integration of interfaces. It also offers message handling services. It was originally reserved for ministries and ministerial offices but is gradually opening up to other bodies or institutions, essentially in the public sector: para-public agencies, political parties, social organizations (trade-unions, etc.), press, Community, Regional, Provincial and Local Authorities, etc. As of now, 900 or so code numbers have been granted, but one code can, of course, be used by several users. Half of the code numbers correspond to effective users of the system.

The different types of services currently available are:

- press agency news and archives of the Belga press agency
- socio-economic databases: Budgetex, Hermes, economic indicators, etc.
- legal databases: Justel, Credoc, Celex, etc.
- specific BISTEL information (national institutions, line-up of the houses, addresses of the ministries, Court of Arbitration, European agencies, BISTEL index , etc.)
- message handling and telex services.

The average monthly length of calls per code number is around 210 minutes, with 16 calls per month on average (May 1990). The most often consulted databases by far are the press agencies (74% of call demands).

B. SECOND WAVE: 1989-1990

Meanwhile, faced with the unexpected performances of the French Minitel, some questions arose in both the public and private sectors (regional authorities, supervising ministries; and software houses, manufacturers, etc) in Belgium and abroad: why does it succeed in France and nowhere else? It is the beginning of what may be called a "second wave" during which several initiatives will be launched as well by public partners as by private ones, aiming at a large-scale development of videotex in Belgium:

- pilot technological sites

    First, at the end of 1988, it was decided by the Ministry for PTT to establish a new generation of VAPs with improved technical functionalities in Liège and Limburg provinces (on each side of the linguistic border): support of the German BTX norm, multispeed modems, download possibilities, etc.

- Videotex Promotion Plan

  This plan was another initiative coming from the Ministry for PTT. Since February 1990, users from five professional sectors (medicine, transport, banking and insurance, etc.) have been allowed to benefit from a monthly traffic credit when acquiring new terminal equipment. This in fact means that the next 2,000 subscribers from the five sectors identified by RTT can see their first videotex bill reduced by as much as BEF 18,000 (ECU 430) in certain cases.

- Termcom

  RTT and other public (Post Office, regional investment companies) and private actors (the five largest banks in Belgium, several terminal manufacturers, etc.) joined in with a "study group" created in May 1989, whose object was to study the possibility of elaborating business plans in order to develop videotex on a larger scale. Some conclusions of this group — inspired by different initiatives launched in the Netherlands— recommended a massive distribution of 100,000 or so terminals which could be rented or sold at moderate charges.

- scattered private initiatives

  Parallel to these "institutional" initiatives, a growing number of private projects arose in Belgium, trying to resolve the chicken-and-egg dilemma on either one side —the terminals— or the other —an attractive service offer. On the one hand, since the opening of the so-called "Gateway" in 1989 allowing the Belgian videotex user to be connected to most of the French Teletel services, different actors (software houses, manufacturers, banks, etc.) proposed that the potential user rented a Minitel terminal (or a modem-card equipment)[47].

  On the other hand, several initiatives, launched by newspapers, local authorities (for instance in the province of Hainaut), non-profit associations, etc., were focused on a limited range of services (in the field of leisure, job vacations, hotel accommodations, etc.), set up on private hosts and to which you can access by means of a password or a specific code number via the telephone network.

Beyond the fact that it is hard to determine which philosophy underlies the choice and implementation of these different projects, one has to admit that none of them has so far yielded satisfactory results.

The pilot technological sites, which had attracted great attention, were displaced and finally their implementation was postponed. What is happening now is a simple technical test of a new generation of VAPs, operated jointly with a manufacturer.

---

[47] At the end of 1990, 20% of the total traffic registered on Minitelnet (gateways between Teletel and foreign videotex systems) came from Belgium: it means about 30,000 hours per year, or 12% of the traffic registered on the Belgian network. One must add that different French service providers have transplanted their services in Belgium so that the traffic related to Teletel may be estimated at about 20% of the total traffic on the Belgian network.

The Videotex Promotion Plan came to nothing: despite the advantages involved (free traffic), the objective of 2,000 new subscribers could not be reached. Even worse, the implementation of this plan corresponded to the period when the increase in the number of subscribers slowed down most spectacularly.

The Termcom study group published its conclusions, but was not in fact able to mobilize the different actors involved in the field of videotex around a common and viable plan.

Finally, the various initiatives trying to encourage the acquisition of terminals have not lead to a significant growth of videotex subscribers, despite the important advertising efforts devoted to them. This is, in fact, not surprising since the situation remains unchanged for the end-users: beyond the fact that the terminals must be rented at market prices, they are still submitted to the general conditions of the RTT system, with the previous subscription, a rather high cost per minute, a lack of attractive national services, etc. As far as the projects operated by different service providers on the PSTN are concerned, the problem is inverse: given the lack of terminals, the service providers are not encouraged to commit themselves on a significant development of their available services; moreover, many of these services are of local interest and cannot be considered as a serious basis for convincing future videotex consumers.

C. THIRD WAVE? THE CASE OF RITT (1991)

During 1991, RTT launched a new project, called RiTT. Its philosophy reflects a reorientation of the public operator's videotex strategy, perhaps due to the proximity of the Single European Market in which value added services —such as videotex— will be opened to competition. The development of RiTT is based on the conviction that videotex, in its classical version, has an essentially professional vocation, unless a strongly voluntarist intervention is undertaken by the public authorities. But everyone knows that in France, such an intervention was motivated by economic and industrial interests. In Belgium, on the contrary, the lack of autonomous national "champions" in the field of telecommunications and the narrowness of the market make this voluntarist option somewhat unrealistic.

In this (new) perspective, the videotex system launched by RTT in 1986 will remain focused on a specific kind of users and its growth will be much more limited than in the initial plans.

The RiTT project is oriented to another target: the domestic users. Based on a Dutch technology, it may be viewed as an "hybrid" system: it enables the cable TV network subscribers of two cities (Mons and Antwerp) to access videotex services through simultaneous use of their TV and telephone sets. Downgoing information uses the cable TV networks and upgoing information is channelled through the networks operated by RTT (PSTN and DCS). The advantage of this system over classical videotex is that it requires no special equipment from the user, except a press-button telephone set (DTMF) and a CEEFAX teletext decoder adapted to the TV set.

Given the importance of the Belgian park of teletext decoders (more than 700,000) and the percentage of TV sets connected to the cable networks (97%), the final target here is very large. Following RTT, 7 to 10,000 households are now potential recipients in Mons and between 70 and 80,000 in Antwerp[48].

---

[48] The teletext decoders are much more numerous in Flanders than in Wallonia, due to the different policies adopted by the two public broadcasting institutes in charge of teletext in Belgium. The Flemish institute (BRTN) chose the CEEFAX norm, which allowed the Flemish viewer not only to consult the BRTN

In each case, three different services are offered: the Magazine (a traditional teletext programme), the so-called "Kiosque" (offering an interactive access to more than 40 services (see table 2) and the Videotex (in fact the RTT videotex system).

Table 2. Categories of services offered on the RiTT "Kiosque"

Advertising	9
Tourism & leisure	7
Radio & TV time-tables	4
Economic information	4
Administrative & legal information	4
Games	3
Transportation time-tables	3
Electronic directories (telephone, fax & telex)	3
Second-hand market (cars, PC's, etc.)	2
Internal RTT information	2
Job vacations	1
Message handling	1
Weather	1
Diverse	3
TOTAL	47

(Source : RTT, Videotex Department)

Access to the Magazine is free. For the "Kiosque", the user has to pay a fee corresponding to an interzonal call (BEF 6 or ECU 0.14 every 40 seconds), without subscription, while the service provider does not receive any income. The case of Videotex is well-known: the user must pay a previous subscription fee, his connection time and a potential entrance fee to the service that he has chosen.

For the moment, all paying services are located on the Videotex. But one of the next steps of RiTT, planned for the beginning of 1992, is to open the access to all services via a single call number (with a multilevel tariff structure, detailed on a separated bill as in France, and a pay-off for every service provider). Clearly, this means that the subscription previously necessary, would be abandoned: a considerable development!

Moreover, RTT intends to market, before the end of 1991, a new generation of telephone sets, with alphanumeric keyboard, enabling the user to create not only digital but also textual information. These new telephone sets will be rented at the same price as the current ones.

Given that no specific equipment is required from the user and that the existing videotex network infrastructure can be adapted without any technical and financial problems, RTT appears to be very confident about the future, in spite of the rather disappointing results of similar projects registered abroad, particularly in the Netherlands. In the latter case, however, the technology

---

teletext programmes but also the British, Dutch and German ones. The French-speaking institute (RTBF) chose the French ANTIOPE norm, much more expensive and unadapted to the existing available services. After several years of total disaster, RTBF decided to abandon its teletext experience. Therefore, of the 700,000 or so decoders set up in Belgium, 95% are CEEFAX decoders, which explains why they are mainly concentrated in the North part of the country (Pichault, 1988).

employed was much more complex and expensive. Unfortunately, no information is available about the traffic on RiTT: either because they are too recent (the two projects started in May and June 1991) or, more probably, the first performances registered are too low to be published.

Videotex: a mirror of RTT's policy in the NIT sector

After reviewing the Belgian videotex situation briefly, one must admit that until now it has developed rather unsatisfactorily. And the question is: is the present lack of success of videotex in Belgium and the absence of any encouraging perspective due to an accumulation of sporadic and individual mistakes, or is it a consequence of deeper, structural causes?

In our view, the essential cause of this phenomenon lies in the particular context surrounding any project in the NIT sector in Belgium. If, as is usually said, a crisis situation is defined by the disappearance of former models, without any new ones being as yet available to take their place, one has to conclude that Belgium is, in this field, in a crisis situation. Indeed, the overall context is dominated at the same time:

- by the phasing out of the old European order in telecommunications, characterized by the monopoly of public operating companies;
- and by the absence of any of the benefits related to the appearance of another model in which new actors would play a part alongside the public companies.

'Neither monopoly nor competition', is the situation in which the new services have to develop in most European countries and the differences that can be noticed from country to country probably originate in the RTT operators' relative resistance to change, in the dynamism of new private actors or, in any case, in their ability to properly manage the transition period.

Neither monopoly

Undoubtedly, videotex is not one of the reserved telecommunication services, likely to be operated in future under monopoly. According to the European directives, not only the videotex services, but also the networks — VAPs, links between them and the host computers and also, if need be, network management centres — should be open to competition. Thus sooner or later, the conditions for a really competitive environment will have to be met, particularly by suppressing cross-subsidization from reserved services.

The Belgian RTT — which remains the main actor in the field of videotex — deemed it wise to anticipate the next inescapable move by granting its videotex department full autonomy and by turning it into a "profit centre" - just as it did for other services like the telephone kiosque.

In fact, the RTT Videotex Department is now in the same position as any potential new operator, having to yield its own margins and being charged at market prices for the services provided by other RTT departments (telephone for the access to VAPs, DCS for use of the X25 network). This is also the reason why RTT charges videotex services and telephone calls separately. It is not surprising, therefore, that the Belgian tariff structure is much less favorable for users than that of most neighboring countries.

As early as 1986, RTT insisted that it renounced any cross-subsidization between reserved services and videotex —which was not the case in France for the Teletel Programme. The objective declared for this activity was to break even after 4 years..

Today, these initial objectives are still very remote, which is not surprising if one considers that, under such conditions, no European videotex service would be profitable. The implementation of a videotex network is a long-term, risky investment, which can become profitable only when the marginal cost ratings of the other services are provided by the operating company, and when all generated external effects are taken into account[49].

In fact, the approach chosen by the Belgian operating company has made any voluntaristic strategy almost impossible and has hampered the development of the service.

If videotex needs a monopolistic environment to develop, this is not only because there is no strong pre-existing demand for it, as opposed to mobile telephone and fax, but also because the development of the network requires a coordination of the initiatives of different actors (manufacturers, network operators, service providers, etc.) and the implementation of a regulation mode that can ensure normal conditions of profitability at each level in the chain. This regulating or integrating function, which is all the more difficult as the market is limited, can be considered as falling typically within the competence of public authorities, although not necessarily (Vedel, 1989, pp.24-26).

RTT has remained caught in an essentially technical corporate culture but it also pursues short-term profitability objectives, for a number of reasons listed above. Consequently, it has proved unable to take into account the constraints of the other actors in the network, particularly those of the information providers[50].

Nor competition

Thus videotex could not rely on the comfortable, and not necessarily constraining structure of a monopoly. Does this mean that it has been able to benefit from the possible advantages of competition? The answer is no.

First of all, the very idea of competition in this field is unrealistic: the short-term profitability of a videotex network being what it is, the number of potential actors is necessarily limited. Nationally, it can only concern the suppliers and/or major clients of RTT, whose interest is to maintain good relations with it, rather than participate in a high-risk sector: one can mention Bell-Alcatel, a subsidiary of the well-known multinational French company which receives more than 75% of RTT's total orders. Besides, foreign actors, like France Telecom, seem to be tied by some code of fair practices between public operating companies that tend to protect their own private territory.

Moreover, if competition was to be considered as the new mode of regulation of the sector, this would imply that all actors — including the public ones — should be motivated solely by economic factors. It is a well-known fact that, when defining its strategies, a public operator like

---

[49] See, on this matter, the debates turning around the profitability of the French Teletel Programme: for instance, Abadie (1989) or Minon & Pichault (1988a).

[50] A similar scenario may be observed for different new IT services, including ISDN (Minon & Pichault, 1988b). In contrast to the partnership policy adopted by France Telecom, RTT preferred to launch a technical experiment and opened it to commercial uses in 1989, without any analysis of the possible needs, the kind of services likely to be offered, etc. The results are naturally not very convincing, with only 350 or so subscribers in 1991.

RTT has to take into account requirements of industrial policy, political constraints or regional interests:

- More than ever, there remains in Belgium a deliberate policy to support national industry even if this entails extra costs or questionable technical choices in preference to the needs of the market.

- One must add that for videotex, as for any other telecommunications service, RTT's choices seem to be strongly affected by political motives. The large number of un-coordinated initiatives (pilot technological sites, Videotex Promotion Plan, Termcom, RiTT) is probably due also to a rapid succession of supervising ministers. One may also wonder about the impact of purely electoral considerations on decisions to launch diverse "experiments" whose usefulness and profitability were at best questionable.

- Regional disparity is one of the most important features to keep in mind when studying the situation of new IT services in Belgium: it may be seen by looking at the number of subscriptions to ISDN, the number of connections to the packet-switched network, the use of value-added services or the penetration rate of teletext decoders. Two theses are confronted here. The first one considers that the problem has to be envisaged from the point of view of demand. Such a disparity simply reflects the unequal economic development of the three regions: old industry in the South (steel, coalmines, etc.), services in Brussels (bank, insurance, EEC, etc.) and new industry in the North (chemistry, high-tech, etc.). It is not surprising therefore that industries located in Brussels and in the Flanders are characterized by a more intensive use of new IT services. The second thesis refers to the fact that the main RTT's supplier (Alcatel-Bell) is established in the North part of the country and that Flanders' interests are strongly represented within RTT. Consequently, one may wonder to what extent the desire of the Flemish part of the country to strengthen its links with the Netherlands and to differentiate itself from France has influenced, for instance, the option in favour of the Prestel norm (see also the choices made by the Flemish broadcasting institute in the field of teletext).

Future prospects

Will competition dynamize IT services as much as some people think it will? What will happen to new services with slow and uncertain development? Will the sole market law permit the establishment of satisfactory modes of regulation at each level in the chain? Would Transpac or the Teletel Programme have been able to develop at such a pace in a deregulated context?

In any case, nothing is worse than an uncomfortable intermediary situation in which the advantages of the monopoly have disappeared and those of competition have not yet arisen.

In Belgium, the public operator does not appear to believe in the future of videotex, conceived as an uncertain sector, but nonetheless it wishes to take as much advantage as possible from the tools at its disposal - tariff policy, activities still under monopoly, leverage on manufacturers and major clients, etc. - in order to block the development of external initiatives whose possible success it would be unable to control.

A clear example of this situation is given by the RiTT initiative. The latter is a subtle solution, enabling the public operator to be present in the field of domestic IT services - despite the various criticisms made of its videotex system - without having to support excessive financial charges. On the one hand, after having, until now, prevented the cable TV operators offering interactive applications (given its monopoly on the transportation function), RTT exploits a legal possibility which allows it to requisition the cable networks for its own purposes (see Pichault, 1985, pp.43-48). It is important to note that, even though Belgium is one of the countries with the highest density of cable TV penetration in the world (97% of TV sets are connected to the cable), the development of these networks has nothing to do with RTT's policy. In fact, the ownership structure of the cable TV operators is largely dominated by private or "mixed" interests, as is shown in table 3.

Table 3. Ownership structure of cable TV companies in Belgium
(in % of subscribers)

- private	33,5%
- public (local authorities)	13,5%
- mixed (local authorities + private partners)	53%

(Source: Deltenre, 1990, p.130)

In other words, cable TV networks are not RTT's property but legally the public operator can make use of them if necessary[51].

On the other hand, until the opening of the terminals market to competition within the Single Market, the first telephone set connected to the network is still under RTT's monopoly. Therefore, in an attempt to take advantage of this situation to support the RiTT project, the public operator will offer to its subscribers, before the end of 1991, a new generation of telephone sets with alphanumeric keyboard at a very attractive price (lower than the actual cost!).

By doing so, RTT wants to confirm its presence on the "general public" videotex scene, without taking major risks: the necessary adaptations of the network are marginal, the financial charges for the terminals will be very low (the extra cost taken in charge by RTT will not exceed 10%) and the information providers offering their services on the Kiosque will do so without charge ... The latter point helps us to understand why so many services presently offered on the Kiosque (see table 2) may not be really considered as "value added": advertising, tourism or administrative information, radio & TV time-tables, etc., ... all kind of "cold" and downgoing information which already proved to be one of the main factors leading to failure the first videotex trials in the beginning of the eighties in Europe (Pichault, 1987).

But it is not certain that the RTT's initiative will modify the main features of the Belgian videotex situation. Indeed, the narrowness of the market, the lack of stimulation for service providers combined with the necessarily higher costs of a bilingual offer, the complexity of the public decision-making system in the field of telecommunications (with different levels of competence intertwined: national government, regional administrations, cultural communities and even local authorities) and above all the strong "engineers" culture and the old bureaucratic structures of RTT question any attempt at developing the IT sector in Belgium.

The RiTT project must be assessed within this context: the two pilot zones are very limited, the information providers may not make money on the Kiosque and then develop the kind of services which do not seem particularly attractive: regional and local authorities as well as the cable

---

[51] The future of this legal framework (after 1993) is not yet clearly determined.

operators complain about not having been consulted before the beginning of the experiment. The specificity of the RiTT project is undoubtedly its technical originality (the "hybrid" way) but one has to admit that it was launched with few considerations for the social and economic conditions of success, i.e. the level of possible demand, the existing park of teletext decoders (probably overestimated in Wallonia), the motivation and the capacity of service providers to develop an attractive offer, etc.

However, the whole history of Teletel in France shows that such elements, not purely technical, were of primary importance in the success of the operation. Therefore RTT should perhaps abandon its traditional role of simple "carrier" and transform itself into a genuine actor within the moving field of communication, where, above all, both commercial know-how and anticipation of social uses are needed above all (see Charon, 1989). The present transformation of RTT into Belgacom - a semi-autonomous structure, like the Dutch PTT - the recent nomination of a new CEO, and the proximity of the Single Market will perhaps stimulate these necessary adaptations.

References

Abadie M. (1989). La stratégie secrète de Télétel, **Télématique Magazine**, n°36, octobre, pp.22-25.

Charon J.M. (1989). France Telecom: un opérateur de réseau devient un acteur de la communication. **Dix ans de vidéotex**, special issue of the TIS and **Réseaux** Journals, vol.2/1—n°37, pp.33-52.

Deltenre Ch. (1990). Le marché belge de la télédistribution — situation et défis", Eurocommunication Research Group, **Convergence of Telecommunications and Broadcasting in European Community. National Monographies and European Synthesis**, Report to the EEC-DGXIII, Liège: LENTIC, 236p., pp.122-143.

Minon M. & Pichault F. (1988a). **De nouvelles opportunités pour le vidéotex destiné au grand public en Région Wallonne**, Report to the Walloon Ministry for New Technologies.Liège: LENTIC.

Minon M. & Pichault F. (1988b). ISDN in Belgium: the State of the Art. In J.Arlandis, **Introduction du RNIS en Europe**, IDATE, 2 volumes, vol.II: pp.1-32

Minon M. & Pichault F. (1989). **Pour une télématique de service public**, Proceedings of the Conference organized by the Fondation Roi Baudouin, January 1989, Brussels.

Pichault F. (1985). La télématique dans le cadre réglementaire et institutionnel de la Belgique, **Courrier Hebdomadaire du C.R.I.S.P.**, n°1101-1102, 6 December, 62p.

Pichault F. (1987). Social experiments with IT from the Initiators Point of View. In L.Qvortrup, Cl.Ancelin, J.Frawley, J.Hartley, F.Pichault & P.Pop, **Social Experiments with Information Technology and the Challenges of Innovation**, Dordrecht/Boston/ Lancaster/Tokyo, Reidel, 317p., pp.241-260.

Pichault F. (1988). La télématique et les réseaux de transmission. In Ministère de la Communauté Française/ Edimedia, **Annuaire de l'Audiovisuel 1988-1989**, Brussels, 385p., pp.109-123.

Vedel T. (1989)
Télématique et configuration d'acteurs: une perspective européenne. In **Dix ans de vidéotex, special issue of the TIS and Réseaux Journals**, vol.2/1—n°37, pp.15-32.

CHAPTER 8

DENMARK: FROM ELECTRONIC PICTURE BOOK TO NEW MEDIUM?

Mads Christoffersen
Institute of Social Sciences
Technical University of Denmark

Introduction

The development of videotex in Denmark has many characteristics of a failure. Both in the opinion of the information suppliers and users and from an international perspective, performance of the Danish videotex system is very poor. This rather depressing situation is not, however, due to a lack of initiative and action. Quite on the contrary: hundreds of millions of Danish kroner have been spent to buy, install, experiment and modify quite advanced technical equipment and thousands of working hours have been devoted to refining this system that for a long period was seen as **the** new communication system that was to pave the road to the glittering information society for the average citizen.

But up until now only a little molehill has been produced from the mountain of effort and investment in the videotex infrastructure. What explanations can be given to this situation? Why do the considerable amount of resources invested in videotex only bring about this limited result? Can it be explained with reference to the characteristics of the actors engaged in the development process or is it due to structural elements in Danish society?

This paper intends to give an overview of the development of videotex in Denmark. In the first section a short history of videotex is outlined and thereafter a description of the current system is given. Then the main actors and phases of the development are presented and in the last section an interpretation is offered. It is suggested that the development of videotex can be distinguished by three different phases and that these phases can be characterized with respect to the specific articulation of a) the dominating socio-technical concept, 2) the driving social agent(s), and 3) the predominant regulatory regime. It is furthermore the contention that the specific combinations of these elements in the current phase to a large extent explains the poor results attained by videotex in Denmark.

A Short history of videotex in Denmark

The introduction of videotex in Denmark has been discussed since the late 1970s as in other Western European countries. The advent of new technologies such as teletext and videotex provoked the formation of a government commission[52] in 1978 that was assigned the task of assessing the effects these new technologies would have on the general public in terms of service provision and everyday life.

In this commission technicians from the PTT, the regional telephone companies (telcos[53],

---

[52] The Commission of Teletext Systems (Teletekstudvalget) was founded on the initiative of the Ministry of State and published its report in 1981 (Teletekstsystemer, Betænkning nr 935).

[53] In Denmark there is a long tradition for regional telephone companies and a central PTT. The regional telcos are KTAS (the Copenhagen area), JT (the main part of Jutland), FT (Funen) and TS (the southern part of Jutland). KTAS and JT are by far the dominating companies - TS, formed in 1987, was earlier a part of the state-owned Telecom Denmark which is responsible for the interregional and international communications. In November 1990, however, a new law on telecommunication fundamentally changed the telecom sector. All the regional companies and Telecom will become subsidiaries of the major holding

H. Bouwman and M. Christoffersen (eds.), Relaunching Videotex, 99–111.
© 1992 *Kluwer Academic Publishers.*

representatives from the press and the state bureaucracy tried to sort out the possible social, economic and political aspects of the introduction of one or several of such systems. The commission agreed that a teletext system similar to British systems (Oracle and Ceefax) should be implemented in the immediate future and could be run by the national broadcaster, Denmarks Radio.

However, the Commission did not recommend the launching of a videotex system due to the uncertainty about both the future role and impact of the new medium and the economic implications. Instead the decision was postponed until a field trial could be established to gain concrete experience.

Underlying this position was a more fundamental conflict of interest between the telcos' eagerness to engage in a new field of enterprise and the PTT central administration representing the general interest of controlling the development of a new medium[54].

The former wanted to expand in an area that looked no different from other forms of modern data communication and was open to competition from alternative system providers, and the latter wanted to evaluate the different aspects and consequences of the new medium before its launching. The administration thus conceived of videotex in terms similar to those established in mass media and considered it convenient that this new electronic medium became an object of the public debate and scrutiny.

When the Danish Media Commission was established in 1980 to examine current media developments and to propose a proper course of action, it seemed obvious to analyze videotex. Consequently a special committee on videotex - in Denmark known as 'TELEDATA' - was formed with the purpose of following the proposed field study and to outline a possible policy for the parliament and the government.

The trials started in 1982 and comprised 100 TV sets as terminals equipped with a videotex and teletext decoder. These were located in private households in two towns: in Lyngby, a northern suburb of Copenhagen and in Silkeborg, a town with 36.000 inhabitants in Jutland. This design of the trial shows the early perception of videotex as a medium primarily aimed at private households as users. However, another 75 terminals were placed in professional environments and further 25 terminals were assigned to public institutions, e.g. libraries and schools.

The use of the TV set as terminal was parallel to the first British Prestel model and the terminals were equipped with full alpha-numeric keyboards. The TV sets, with built-in videotex/teletext decoders, were ordered from the Danish hifi producer Bang & Olufsen.

The accessible information was provided by travel agencies, the press, publishers, banks and various other organizations taking part in the trial. The system was based on the British Prestel standard and paralleled the original Prestel architecture with two central computers, one in Copenhagen and one in Århus serving the whole system. These were built by the Danish electronic firm Christian Rovsing which was strongly expanding in various data processing areas throughout the late 1970s and the early 1980s.

The first system was owned and run by the four Danish telcos and the PTT as a joint initiative. KTAS and Jutland Telephone, traditional rivals - although not very often in open competition due to the partition of monopoly concession areas - were the dominant players and each insisted on

---

company, Tele Danmark, which will operate as an independent agent although 51% of the stocks will be in the possession of the State. From January 1991 South Jutland Telephone, Funen Telephone and Telecom have belonged to Tele Danmark - KTAS and Jutland Telephone will enter the group in March 1992, cf. Abild Andersen, 1991.

54   Until 1986 the operating branch of the Danish PTT (Statens Teletjeneste) has a formal monopoly on data communication. Although the monopoly was gradually undermined by the digitalization of the regional companies' network both the PTT and the companies wanted to position themselves with regard to the new data communication services.

having a machine centre. The two centers were thus constructed with equal ranking in the hierarchy, a peculiar architecture causing many problems with updating of the information.
Industrial policy was reflected in the fact that both the terminals and the central equipment were produced by national firms (Bang & Olufsen and Rovsing) interested in developing competence within a promising new industrial field.

From a technical viewpoint, the system was provided with a rather advanced facility permitting alpha-numerical retrieval of all information. The results from the trial over a two year period from April 1982 to April 1984 were, however, rather ambiguous.
The trial was closely followed by a group of researchers and they concluded in their final report that the information provided by the system was lacking specific address. With the exception of the electronic phone book and the system's rather large encyclopedia, it was hard for the users to see the utility of the stored information.
The users did not find the TV set to be an appropriate terminal for videotex. It sometimes conflicted with the interest of watching television and many found it hard and tiresome to perform the possible information retrieval.
The reactions from the information providers was also less than enthusiastic after the trials. The opinions expressed by the Press Association in a concluding report was symptomatic of the general impression: videotex didn't seem to be a "field of appropriate economic interest for the present time".
Other information providers were likewise disappointed. Several banks had participated in the trial and their experience did not encourage them to further engagements: there were numerous technical problems with connecting the systems, the telephone companies were conceived as inflexible towards the IP's needs and the customer base did not justify larger investments.
After the end of the trial there was an interim period of one year before the Media Commission submitted its final report in 1985. It contained a special chapter on videotex (Teledata) in which the experiences from the field trial were summarised. The Commission regarded videotex as a technology primarily aimed at private users and supported the idea of establishing a public videotex system to be run by the telephone companies. On the other hand, it seems obvious to the commission that the videotex network and services could not be reserved as a part of the telecom monopoly. The argument was that there was no fundamental difference between the materiel used for videotex and that used for private datacommunication.
The consequences that would result from the development of several different networks without any coordination and regulation worried the commission. Therefore the importance of developing of a public network was stressed but no specific proposals as to how this could be done were presented.
Consequently, the Media Commission did not put forward any consistent or precise policy concerning the launching or use of the medium. A minority of the Commission expressed the view that videotex systems were only of societal interest as long as they were "formed as a contribution to the information society from which many groups of the population can benefit"[55]. As far as the question of advertising was concerned, it was the general opinion that videotex should be subject to this type of information. However, a group of members voiced the view that adverts should be designed in such a way that there was no doubt as to the nature of the message. This proposal has not been supported by specific regulation up to now.
Although the results of the trial were somewhat unclear, there was a certain optimism among politicians and leading telecommunication managers about the future of the new medium. They strongly believed that videotex would quickly gain in popularity and it was assumed that, by 1990, there would be about 100.000 subscribers.

---

[55] The Danish Media Commission (Mediekommissionen): Final report p.: 186. Betænkning om dansk mediepolitik, nr 1029, 1985.

The commercial phase of the first videotex system in Denmark began in the worst possible conditions. In the last year of the trial, the Danish manufacturer of the system, Christian Rovsing, went bankrupt, and his enterprises were parcelled out and sold.

One of the consequences of this disruption, was that the telephone companies were left with a system without the necessary documentation. The technicians responsible for the hardware and software moved on to other companies and others went abroad.

Under these conditions, the companies had no possibility of actively marketing the new service. In the years 1985 to 1987 the number of subscribers never exceeded 1400 and many of these were the telecom administrations' own employees who wanted to try out the possibilities of the new medium.

The absence of a genuine take-off of the service gradually disencouraged the information providers. Some withdrew from the service, others neglected to up-date their information, and many complained about the bad service and lack of results with regard to promised improvements. So all in all, the initial interest from information providers gradually cooled during the trial and it almost disappeared in the following years.

Although the central information provision in the two centers were supplemented by a x.21 gateway solution for external computers, potential information providers gradually endorsed a more decentralised solution.

To summarize the resulting lessons of the first videotex system in Denmark it can be stated that:
- using the TV-set as terminal was not an obvious success among the users[56],
- it was not made clear that private users were the main target group for the new medium,
- the centralized system architecture caused discontent among information providers as they became aware of the prospects of operation of their own machines,
- it turned out to be a hazardous undertaking to rely on a national electronic producer as the hardware provider, and therefore,
- videotex did not seem suited for conducting a national industrial policy in Denmark.

The current videotex situation

When the decision was taken in the mid 1980s by the directors of the regional telcos to launch a new system, a fundamental intention was to avoid the failures of the old generation. This applied to the target customers, system architecture and standards.

As opposed to the first generation, the new system was supposed to focus primarily on the professional customers and not on residential users. The latter would be targets for marketing in a later phase when business services were more thoroughly developed.

Instead of forcing a new terminal or communication protocol on the business customers, it was believed that the "terminal problem" could be partly solved by basing the system on the standards of already existing terminals and on the multi-user systems.

Given this background it was essential to create a system that could use not only the bulk of videotex standards established on the European market since the beginning of the 1980s, but also the widely spread TTY-terminals applied in the many multi-user systems in the industries and administrations. The fundamental concept being that the user, should, regardless of his terminal,

---

[56] For a more comprehensive account of the relationship between broadcasting and telecommunications in Denmark see: Christoffersen & Skouby (1991).

through the system have access to all types of services and bases[57]

This multi-standard architecture is conveyed through a **transcoding** facility located in the videotex access points (VAP).

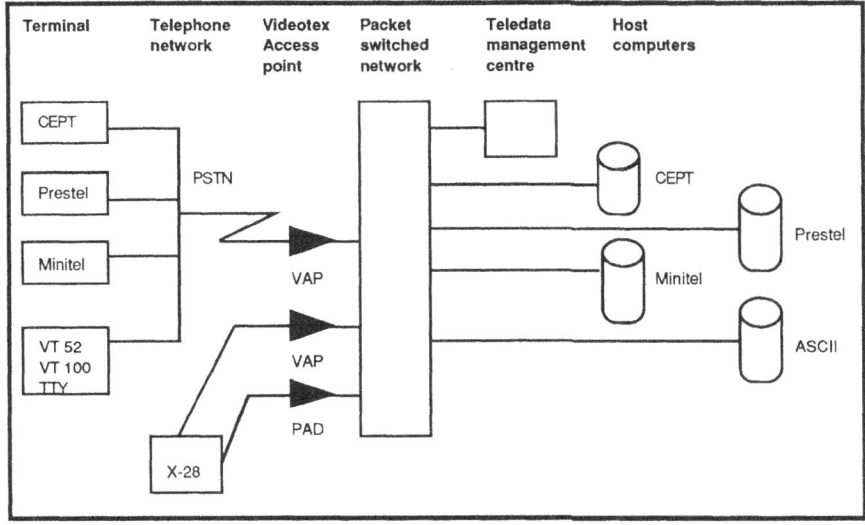

Figure 1: The architecture of the Danish Teledata system

With this architecture the information providers have the choice of either having their machines connected directly to the packet switched network or finding an appropriate host for the service. Both major regional telcos in Denmark (KTAS and JT) have facilities to function as hosts for videotex service providers in Denmark.

As far as the **supply side** is concerned, the Danish Teledata offers connections to an impressive number of services. Until the beginning of 1991 around 1550 different bases and services could be consulted. In January 1991 the Danish telephone companies signed a contract with France Télécom that opened access to the entire Télétel network with more than 16.000 services.

But apart from this French connection it must be kept in mind that the vast majority of accessible service are not videotex services in the traditional sense. They are international ASCII bases[58] reached by the Teledata network as an alternative to the X.25 network or circuit switched network. In this sense the Teledata system functions more as a switchboard for the information retrieval users than as a genuine videotex system.

The bases registered in June 1991 were distributed among 100 different hosts, and the vast

---

[57] Despite this principle the system did not include the French Télétel standard from the beginning. This shortcoming was corrected in August 1990 where it was made possible to access Tététel bases with Minitel-terminals. At a later date it will be possible to connect to the system with other terminals.

[58] Among these are international information providers as BRS, CompuServe, Data-Star, Dialog, Dimdi, Echo, ESA-IRS, Eurobase, Finsbury, Fiz Technik, Kompass, Pergamon, Profile, STN, Questel. All of these are costly professional databases requiring knowledge of their, often highly specialised, search language.

majority of these are international services located in foreign countries. Only a minority are open to the public without subscription:

Table 1: Danish videotex and information services on Teledata

	Open bases	Closed bases	Total
TTY/ASCII	43	1454	1497
CEPT	10	0	10
Prestel	23	16	39
Multistandard	10	0	10

Source: KTAS Teledata

This reduces the supply of Danish proper videotex services to 78 of which some are reserved for closed user groups. The rest includes some minor banks, some interest organizations, some business organizations, travel agencies, some specialised information services and a couple of statistical services. It is furthermore possible from the Teledata system to use the x.400 electronic mail system and to send telefax.

In August 1991 the three most used services was: 1) a credit information service - reserved for subscribers within the retail sales sector, 2) the electronic directory, 3) classified ads for selling and baying.

But none of the newspapers or publishing houses that participated in the trial, are represented as information providers. Their attitude towards the Teledata service is rather skeptical and they are clearly, not willing to make investments in further service provision until the nature and form of a potential future market had become more apparent.

As far as the **demand side** is concerned, the system started with a subscriber base of 924 at the opening in April 1988. This number rose to 2005 in April 1989, to 4222 in April 1990 and then to 6379 in April 1991. In September 1991 the system had 6625 subscribers and the figure fell to 6301 in October. This seem to indicate that a peak has been reached within the present service configuration and market strategy.

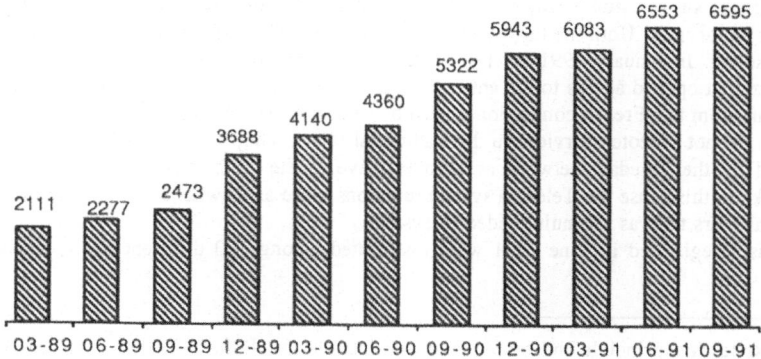

Figure 2: Number of subscribers to Teledata April 1989 - October 1991.

Source: The KTAS and the Teledata system.

A large majority of the subscribers are business customers from the sectors shown in table 2.

Table 2: Subscribers' distribution on branches of industry

	absolute	percent
Data industry	440	7
Agriculture	52	1
Raw material	10	0
Industries	451	7
Building industries	166	3
Trade & commerce	572	8
Transport	1125	17
Finance	639	9
Services	1462	22
Information providers etc.	260	4
Private households	1425	22
Total	6605	100

Source: Teledata service.

Although some progress can be observed, it is evident that the videotex system of the telephone companies has not developed sufficiently to "take off". Even though private households constitute the second largest user group it lacks the mass that would attract information providers. The total system receives around 55,000 calls per month, and 52,800 connect hours were registered in 1990 according to official statistics[59]. This gives an estimated average annual duration of around 10 hours per user or 50 minutes per month.

This lack of dynamism has influenced the potential information and service providers' willingness to invest in creating videotex services. Although there has been some growth in the number of services available, only a few companies and institutions have invested the necessary resources to make a qualitatively satisfying service with sufficient information and transaction possibilities. This contributes to the impression that Teledata in Denmark is not a **living** medium or a suited marketplace for buying and selling teledistributed services.

The question of tariffs has been a source of some internal disagreement between the regional telephone companies running the Teledata. From the beginning it was clear that an open service provision like the French "kiosque" would not be chosen. Subscription was seen as the only way to separate Plain Old Telephony from videotex and connected value added network services (VANS). There was, however, dispute about the price for subscription. Finally, it was decided not to charge for joining the system but to make the subscribers pay an annual fee of 300 DKR[60] (= 38 ECU).

The traffic tariffs were at first stipulated at a level of 60 DKR pr hour (7.58 ECU). After a period, however, it became clear that this rather high level did not stimulate the development of the service. The tariff was therefore lowered to 45 DKR (5.76 ECU) pr hour for the daytime and 23 DKR (2.88 ECU) for off peak hours and week ends.

Besides the traffic, the Teledata system also contains the possibility of collecting fees for the information providers. Two methods are available: a page fee and a time fee up to 99.99 DKR

---

[59] The management for the Teledata service refuse to publish statistics on traffic arguing that the service is placed within the competition area. Information on the total number of connect hours appear, however, in the Tele Yearbook 1990, p.17. The number of calls is not officially known, but Videotex International no 138/139 1991 mentions 55,000 connections in March 1991.

[60] All tariffs are indicated including V.A.T. and the rate of exchange DKR/ECU is 1:7.91.

(12.6 ECU) per page or per minute.

The tariffs for the information providers consist of different elements. Even though the connection to the network in principle is free for the information providers, they have to pay an initial fee of 500 DKR (63.2 ECU) for the gateway information page and an annual fee of 1000 DKR (126 ECU) and to this adds a traffic-dependant volume tariff[61]. To these costs must be added the cost of a users subscription as each information provider automatically is registered as an ordinary subscriber.

There is no available information on the amount handed over to information providers from the operating companies.

Agents within the videotex field in Denmark

The public Teledata system is, however, not the only system in operation. A handful of networks operate outside the telcos' system using 'dial-up' to their own machines. Most of these networks are specialized applications within the business sector or designed to specific customers needs. The most important are the bank services, running parallel to the leased line solution or the IBM 3270 protocol, very widespread in the Danish banking sector, which is offered to the bigger customers.

A more coherent alternative is provided by the INS service run by IBM in Denmark. The customers of this network dial up to an access point in one of two locations and the traffic is routed through IBM's own network. The service is connected to the general network and supports Prestel, Télétel, Bildschirmtext as well as the TTY protocol and opens up for possibilities of extensive international communication. Information about the price policy and the content and quantity of the service is not available.

Neither of these networks claim to constitute alternatives or to be competitors to the telcos' public Teledata. This, however, is exactly the position of the most recent actor in the game: **NetPlus** - a subsidiary of the municipalities' EDP enterprise.

NETPLUS - A CHALLENGER WITH FRENCH ACCENT

NetPlus was formed in 1988 as a subsidiary of **Kommunedata**, one of the biggest data processing firms in Denmark with an annual turnover of 1 billion DKR (= 126 million ECU). Kommunedata provides the Danish municipalities with data, data processing and services, and is totally owned by the Municipalities Association. It thus operates as a private firm for public institutions with monopoly of 'data catch' in some areas and in competition with private service providers in others.

Kommunedata has for some time considered the possibility of opening a completely new field of service: providing electronic information and services **directly** to the citizens of the municipalities. Videotex was seen as an appropriate technical solution and different national systems in Europe were scrutinized. Different demands were formulated as criterias to determine the appropriate system:

---

[61] This tariff is calculated on the basis af 0.00044 ECU per segment in the packet-switched network.

- the terminal should be cheap
- the system should be easy to use
- the traffic tariffs should be low
- access for users should be open
- access for information providers should be easy.

The only system in Europe fulfilling these demands was the French Télétel system which, by 1987/88, had proved to be a success in terms of user-friendliness and traffic.

NetPlus thus decided, as much as was possible, to import the French model. On the service side the intention was to supply municipal services within for example the social, the health and the tax areas. Furthermore the system was intended to encompass electronic phone books, news, teleshopping, classified ads, games, electronic mail and chatlines modelled on the famous French "messageries rose".

The close relationship between NetPlus and Intelmatique, the France Télécom subsidiary responsible for the marketing the Télétel concept abroad, indicated the possibility of a new French strategic operation.

After the domestic success, France Télécom oriented its policy towards the possibilities of export. Different initiatives were launched in Europe and USA but without the same success as in France. It therefore seemed plausible that France Télécom (Intelmatique) had been considering the possibility of choosing a small European country where it could demonstrate the viability of the "French model" in quite another national environment.

In order to pave the road for the launching of a "French" operation that necessarily would demand the participation of other major players in Denmark - e.g. the telephone companies, financial institutions, the news papers - NetPlus has carried òut a number of field trials.

Three different trials were started in the spring of 1990. One in a minor rural community where 100 households have been equipped with Minitels[62] and smartcard readers enabling the users to access a municipal information base, a teleshopping service and different communication facilities. Another has been opened in a municipality in the outskirts of Copenhagen, Ballerup, where terminals have been set up in public places providing information to the local citizens on housing and employment questions.

A third initiative was taking place in Odense, a town of 175,000 inhabitants, where the local newspaper supplied news, classified ads and games. This system also contained an electronic mail system and a chatline service. The local radio also joined the system as an information provider proposing different entertainment services.

The terminals for the trials were delivered by Philips at a monthly rental basis of 39 DKR (= 5 ECU) with the terminal producer as the financier of the rental operation. The use of the services were charged at a time-based fee of 41.40 DKR (5.20 ECU) per hour which is to added to the cost of a local phone call[63]. 700 terminals have been distributed in 1990, considerably less than the 1000 terminals originally planned. The Odense trail, however, dried up in 18 months because of lack of attractive information. No information or service providers were willing to joint the experiment and make the necessary investments. This trial was therefore together with the trial in the rural area closed down in the summer of 1991 whereas the experiment in Ballerup will continue in 1992.

---

[62] The Minitel terminal is for copyright reasons renamed the "Miniplus" terminal.

[63] From 1.1 to 2.2 ECU per hour depending on distance and time of the day.

OTHER INITIATIVES

In the initial phase NetPlus presented the trials as the beginning of a development leading to a Danish videotex market of 300.000 to 400.000 Minitel terminals in five years. However, it quickly turned out to be totally beyond the capabilities of a municipally based EDP company to install and run a multi-million ECU information network service without the cooperation from the regional telephone companies.

The NetPlus challenge from 1989-90 did bring new dynamism to the development of videotex market strategy in Denmark for a period. Both major regional companies published plans for videotex launches in strategic areas: Jutland Telephone announced the intention to launch a major campaign beginning in the town of Horsens with approx. 46.000 inhabitants. The first phase was market investigation showing that around 25 % of the households declared the willingness to pay for information service after the initial trial year. The Horsens operation was designed to utilise bi-standard terminals (Télétel and Prestel) and was originally planned to take off in 1991. The final decision has been postponed for more than a year and it now seem highly unlikely that this operation will ever take place.

In the spring of 1991 the KTAS announced the intention of launching a similar operation in Lyngby, a high income suburb in the northern part of Copenhagen. This project has identified a target group of about 7000 households that are considered ripe for electronic services as telebanking, local information, electronic directory , electronic mail etc - services to be based on the Télétel protocol and Minitel terminals. But the local company, KTAS, has not yet taken any decisive decision.

Another aspect of the response that was generated by the initiative of NetPlus was the attempt to set up a joint-venture company combining the telephone companies, Kommunedata and PBS (the joint company operating payment transactions in the Danish banking sector). After a long series of negotiations the attempt to create this cross-sector venture was postponed or probably given up due to the difficulties of setting up the functional requirements for a major market initiative. It has been reported that difficulties of coordinating the undertakings of the regional telephone companies played an important role in this abandonment.

All in all the Danish situation can be characterized by a slow and confused development where the supply side has not yet reached the ripeness and abundance that is the indispensable precondition if the users needs are to be met.

This insight was formulated quite succinctly in the summer of 1991 by a director from Jutland Telephone, who bluntly announced that the existing Teledata system was a failure and that the development of a new system 'Info 24' with quicker transmission and shorter response times was under consideration. The new system would be based on the so-called Intelligent Network currently being constructed for tariffing the coming audiotex premium rate services and were supposed to be prepared for the ISDN that will be marketed from the beginning of 1992. Only high quality information providers will be present in the prospected service.

The explanations for this overall rather depressing development of the Danish videotex medium must be sought for at the policy making level, and within the context of the legitimation of speci- fic interests in the conflict-ridden Danish environment. In the last section of this paper I shall try to make a tentative analysis of the participants and their actions.

Agents and contradictions in danish videotex

It seems to me that three different periods can be distinguished in the development of Danish videotex over the last 10 years. The phases differ with respect to a) the **socio-technical concept**, b) the **driving social agent**, and c) the **predominant regulatory regime**. My proposition is that the degree of "success" or degree of matching between a teledistributed service such as videotex and the social responsiveness to it, depends to a large extent on the specific articulation of the three factors mentioned above.

The phases to be described are:
1st phase: **Videotex as the picture book of the national monopoly**;
2nd phase: **Videotex as the multi-standard technical fix**;
3rd phase: **Videotex as a new multi-functional medium**.
As to the concrete periodization it is my allegation that the present period marks the transition from the second to the third phase.

1. VIDEOTEX AS THE PICTURE BOOK OF THE NATIONAL MONOPOLY

This phase marks the very beginning of videotex where faltering innovation resulted in a failed attempt to carry through the compulsory amalgamation of the telephone with the television.
The socio-technical concept of this period is closely connected with the driving social agent, the teletechnical engineers and bureaucrats within the PPT/telcos responsible for launching the new "device". In this period videotex was seen as basically an electronic clone of the printed book or newspaper.
The medium is typically referred to as the electronic "picture book" by the technicians who thought that its most important application would be in areas where the advantages of the electronic vis à vis the printed media would rest on quicker up-dating and faster consulting of information. Parallel with this line of thinking is the technocrats' focus on the possibilities of functional optimization of every day life for the busy yuppy[64].

As far as the regulatory regime is concerned the first phase is dominated by the traditional telephone monopoly that acts as the initiator of innovation. The basic terms of the field trial and provision of equipment is heavily influenced by the interests of industrial policy, thus pushing forward the major Danish terminal and system providers as producers. Another aspect is the rather centralized system which is controlled by the telephone companies.
The **results** of this first phase are generally negative: there appears a conspicuous mismatch between on the one hand the telcos as system operator and carrier, and on the other hand the information providers' and the end users' demands. The general outcome of the first phase is **frustrated aspirations** on the side of practically all important players.

2. VIDEOTEX AS THE MULTI-STANDARD 'TECHNICAL FIX'

The second phase can easily be seen as a reaction to the frustrations of the first. The very closed and narrow socio-technical concept of the first is given up to the benefit of its exact opposite: a videotex system designed to embrace all possible standards. This logic misleads its proponents to simplify the concept of videotex from its role as a genuine medium to a technical gadget. The new Teledata system is technically among the most advanced in Europe, but it does not offer any clear service profile with appeal to information providers and users. The system is caught in a dilemma of wanting to satisfy everyone on the one hand while lacking the ability to satisfy anyone on the other.
This technico-fixed concept stems from the teletechnical engineers who, as in the first phase, play a dominant role. They have carried through an innovation process based on the idea that a new videotex system is viable provided that the technology is sufficiently sophisticated. But technical sophistication is not enough to create a break-through for the new medium. The all embracing switchboard is of no use if there are no services and no subscribers to want them.

---

[64] Cf. the analysis of Jean-Marie Charon, 1987.

In the second phase the engineers within the telco environment play a leading role in concept formulation. But because of the extremely regionalised structure within Danish telecommunications there are serious difficulties of collaboration and coordination - especially between the two major companies. However, the engineers who put forward the idea of videotex services experience ridicule from the agents of the dataprocessing industry. From the latter's viewpoint videotex has every characteristic of a genuine **flop**: it is easy to use, cheap and simple - consequently it lacks all aspects of the information processing set-up that adds weight to the **fanciness** and **mystification** of their profession. Because the yuppy layer of the EDP industry are the opinion-makers of the business, they pay a role as negative agents and create a problem of legitimization for videotex in this phase.

Although the new Teledata system experiences a certain growth in this period, it by no means reaches the famous 'critical mass'. The process towards this goal is furthermore hampered by a shift in the regulatory regime.

The ongoing **deregulation of telecommunication** has given birth to completely new conditions for value added services like Teledata. Tariffs are meant to reflect the "real costs" and no investment in the competition sector can be made with money earned in the monopoly sector. This position radicalizes the questions to the public videotex service from the telco directions and the ruling liberal government: do the earnings from the service justify further investment? In this context it seems hard to legitimize further investments until the service has demonstrated its viability.

The outcome of this phase seems to be a deadlock on a "higher level" than in the previous phase: moderate growth but stagnating investments until the service proves viable. But without further resources nothing will happen. The open acknowledgment of the failure of this concept signals the end of this phase and the possible beginning of an other.

3. VIDEOTEX AS A NEW MULTI-FUNCTIONAL MEDIUM

The current international development of videotex characterized by the apparent French success story and the important investments of IBM and Sears, Roebuck & co in the Prodigy service in the US, indicate a shift in the general view of videotex. The overall image of flop and failure is gradually replaced by the perspective of a new medium with new horizons. Important projects and investments are being prepared or realized in countries as different as Spain, Italy, Ireland and Sweden as part of this reorientation.

Although this tendency can easily be over-interpreted, the process is by no means fully complete. It is my contention that a new socio-technical concept is under construction. This concept reflects more fully the characteristics of videotex as a medium in itself, as a qualitative innovation combining **interaction, communication** and **transaction** with real-time possibilities creating potentialities of new services and functions. The profile of the media-concept is much more marked by the communication aspect than the concepts of the previous phases.

This shift in the understanding of videotex is paralleled by the entrance of new social agents. In Denmark the telephone companies no longer dominate the scene; their understanding of the medium and their market strategy is being challenged by the NetPlus initiative that redefines the battlefield. Other new agents are waiting in the wings. This goes for the financial sector and the newspapers that will probably change the medium in the direction of an **edited entity** - a tendency already observed in France and the US[65].

---

[65]   Cf Pajon: Le vidéotex comme industrie culturelle

Table 3: Main phases in the development of videotex in Denmark.

	Phase 1	Phase 2	Phase 3
Socio-technical concept	picture book	switch board	new medium
Driving social agent	PTT as monopolist	telcos/telcos' engi-neers	telcos, information providers and sy-stem coordinator
Regulatory regime	monopoly	deregulation	coordinated policy?

Although it may be premature to talk about a shift in the regulatory regime in Denmark, there are signs of new initiatives. The realization of the weakness of a totally market-dependent development calls for concrete action in order to overcome the "terminal problem" and for an incentive to activate the hesitant information providers.

Initiatives are planned and the telephone companies are about to give up their reluctance to invest further in the videotex medium. At the same time the conflict-ridden relations between the telcos and NetPlus seems to be gradually replaced by a collaborative discourse forced about by the collective insight that the country is to small to support two competing networks. The newly formed holding company, Tele Danmark, which from March 1992 will control the entire telecom sector, may eventually prove to be the agent that has the sufficient strength and authority to generate a new trend of development and to perform the role as 'system coordinator' much needed in small countries with a weak customer basis.

References:

Andersen, Jørgen Abild (1991). New Telecommunications Structure in Denmark. **Teleteknik,** vol. 34. no 1.

Betænkning nr 935 (1981): **Teletekstsystemer.**

Charon, Jean-Marie (1987). Télétel, de l'interactivité homme/machine à la communication mé-diatisée. In: Marchand, M (ed): **Les paradis informationnel,** Paris.

Christoffersen. M & Skouby, K.E. (1991). The Convergence of Telecommunications and Broadcasting in Denmark. **Communications & Strategies** 4, 1991. pp 105-128.

The Danish Media Commission (Mediekommissionen) (1985). Final report. **Betænkning om dansk mediepolitik.** nr 1029.

Pajon, Patrick (1989). Le vidéotex comme industri culturelle. **TIS** vol 2 no 1/**Réseaux** no. 37.

Tele Yearbook 1990. Statistical Information on Telecommunication Services in Denmark compiled by the National Telecom Agency

Videotex International. no 138/139 1991.

CHAPTER 9

SPAIN: GREAT EXPECTATIONS - A NEW WAVE OF OPTIMISM

Santiago Lorente .
High Technical School of Telecommunications Engineering
Polytechnical University, Madrid

Introduction

This paper[66] is divided into three parts. The first part provides a general description of the present-day situation of Videotex in Spain, "Ibertex", giving summary statistics about users, traffic, number and type of information and equipment providers, network technical characteristics and the main services offered to the users.

The second part presents a brief picture of the development of Videotex in Spain, outlining the main events throughout its short history.

Finally, the third part is an analysis of the policies that are presently being formulated in the wide Videotex scenario, in which four main actors play a role: equipment providers, information providers, network carrier and users.

It is useful to keep in mind that in Spain there exists some linguistic confusion as to the term "Videotex". On some occasions, the term "Videotexto" (Spanish ending) is used and on others, "Videotext" (English ending), but the latter has nothing to do with the German ARD's or ZDF's Teletext. Sometimes it is confused with "Teletexto" (unidirectional information service provided by Spanish Official Television) or with "Teletex" (a service similar to Telex).

General description of the current videotex situation

NUMBER OF USERS, TRAFFIC STATISTICS

Table 1 shows the main parameters related to the evolution of videotex in Spain. Tables have been provided by Telefónica (Spanish Operator). Those given by the Videotex Service Providers (APV) tend to be slightly lower.

---

[66] The following persons have contributed to the realization of this paper:
Mr. José Luis Rebollo, Ibertex Head Officer, Telefónica
Mr. Víctor Domino, General Coordinator, User Association (AUVE)
Mr. César Sanz, Director Generel, Videotex Promoter Association (APV)
Mr. Germán Fernández, Home Banking Director, Banco de Santander (BS)

H. Bouwman and M. Christoffersen (eds.), Relaunching Videotex, 113–123.

Table 1. Main indicators showing spanish videotex evolution

	1988	1989	1990	1991 (june)
Terminal/User	5.000	35.000	70.000	275.000
Service Centers (NRI's)	20	45	200	326
Monthly calls		160.000	387.000	650.000
Monthly connection hours		32.000	65.500	110.000

(NRI = Data Network Access Numbers)

The User Association (AUVE) made a survey (sample n=38,500, which is not representative due to the voluntary nature of respondent participation) and the results pointed out that there are three main user types: 1) young liberal professionals, 2) young office clerks and 3) students. Of particular interest is the fact that 45% of respondents use videotex outside the home, and another 45% use the service in the morning. This phenomenon makes clear that users take advantage of the situation that the use of telephone services at the office costs them nothing. This activity constitutes a kind of "fringe benefit" since there is hardly any control of the use of the telephone in the Spanish offices at this time.

AMOUNT AND TYPE OF INFORMATION PROVIDERS

According to Telefónica (July 1991), there are 326 NRI's (Data Network Access Numbers or simply Network Directions), connected to some 270 external computers which, as a whole, provide more that 3000 different services or applications.

Current data show that 20000 users regularly do food shopping through teleshopping (compared to 3000 in 1989). Some 5000 monthly connections occur. The sociological profile of the teleshopper is: upper and middle-upper social class, 30 to 45 years of age, married, with children aged 18 or less.

SYSTEM/NETWORK: STANDARDS, CONFIGURATION

The Ibertex standard is CEPT-1, Recommendation T/CD-06-01. According to a Telefónica spokesman, this standard has been chosen because the company understands that it is the most complete and future-oriented among the four profiles, and because this standard has been selected by most European countries. This, of course, is subject to controversy.

The elements in the Ibertex chain are: 1) User Terminal (44% on PC; 43% Dedicated; 12% on TV with decodefier); 2) Switching Network (asyncronous, 8-bit, full-duplex, protected, and non protected, CRC error detection); 3) "Iberpac" Network (Spanish Data Package Network running on X-25), 2400, 4800 and 9600 bpi speed transmission; 4) Ibertex Access Center (CAI), which is a virtual center that gives network access to the user; 5) Control and Management Center (where invoicing and statistics are made); 6) Interconnection Videotex Center (for international and other CEPT-standard connections Ibertex Access Centre (virtual network center); 7) Service Center, which provides the information required.

The CAI (number 4 of the above chain) performs several functions and basically makes the dialogue between the user and the service center possible.

COST STRUCTURE: WHO IS PAYING FOR WHAT?

The Service Center pays the tariffs related to the Data Network (Red Iberpac). The various concepts and prices are shown in table 2.

Table 2. Payments to Telefonica due by the service center

	Pesetas	ECU
Enrollment fee	33,200	255.4
Monthly subscription: *** according to transmission period**		
2400 bps	23,355	179.6
4800 bps	36,488	280.7
9600 bps	53,224	409.4
* according to facilities		
Multichannel access. For each additional channel after the first	776	6.0
reverse charge	600	4.6
non-standard package size	600	4.6
non-standard window	600	4.6
multu-link access	1,800	13.8
hunt group	1,880	18.8

Videotex tariffs, approved by Government Law (April 13, 1991) are illustrated in table 3.

Table 3. Payment to Telefonocas kiosk due by the videotex user

Service level	Call Cost Ptas ECU	Cost per minute (Pts/ECU)			Type of service
		For Telefóni ca	For the Service center	Total	
031	11.70 0.09	8.50 0.07	0.00 0.00	8.50 0.07	Level 1
032	11.70 0.09	9.50 0.07	9.50 0.07	19.00 0.15	Level 2
033	11.70 0.09	10.50 0.08	15.50 0.08	26.00 0.20	Level 3
034					Free infor- mation
035	11.70 0.09	3.5 0.03	0.00 0.00	3.50 0.03	Collect call
a) 036 b)	11.70 0.09 11.70 0.09	50.00 0.38 =50.00 0.38	0.00 0.00 Variable	50.00 0.38 Variable	Access to inter- national videotex

SERVICE OFFERED: AMOUNT AND TYPE

According to Telefónica, the main, first-level classification of Information providers is:

1)   **Commerce**: Fairs, International Trade, Shops, Consumers, Trade Laws, Credit Cards, Teleshopping.

2)   **Economy and Employment**: Agriculture, Banks, Stock Exchange, Statistics, Fairs and Congresses, Industries, Services, Jobs.

3)   **Education and Culture**: Arts, Libraries, Movies, Theaters, Discs, Books, Teaching, Museums, Press, Tele-education.

4)   **General Information**: Sports, Economy, National News, Weather.

5)   **Computer and Telecommunications**: Communications, Electronics, Electronic Guides, Computer, Telematic Services, Telematics (C & C).

6)   **Social and Political Institutions**: Government, Foundations, Parliament.

7)   **Messages, Home, Leisure Time**: Hobbies, Restaurants, Horoscopes, Games,

Messages, Tele-software.

8) **Social Services**: Consulting, Address, Health and Medicine, Religion and Society, Public Services, First Aid.

9) **Travel and Tourism**: Customs, Ports and Airports, Lodging, Festivals, Gastronomy, Touristic Routes, Transport, Travel.

Services most commonly used are those which are more interactive, short and frequent: general information, transactional, specialized information and communications services.

Short national history of videotex
Development - main events

In 1971 the Spanish Package-switched Data Network (RETD), the world's first public and commercial network of its kind became fully operational. This event constituted the technological bases for videotex, which was born on an experimental basis in 1977. In 1978, the Government granted Telefónica the exploitation of Videotex. CEPT-1 emerged in Germany, and Spain adopted that standard. By 1979 the Experimental Videotex became commercial. During the World Championship Football in 1982, videotex, though still experimental, became fully operational. In the same year, the Videotex Service Provider Association (APV) was born (I Videotex Congress). In 1986 a new impetus was introduced in the form of General Elections. The II National Videotex Congress took place.

In 1987, several access levels (03x) were created, and Videotex was officially inaugurated. The 10-year lag was due to the unwillingness on the part of the Industry Ministry to homologate Videotex terminals. Such unwillingness must be understood in terms of a permanent conflict of interests between both Ministries in controlling the powerful telephone company.

Invoicing becomes more transparent: costs are charged to the user meter and are computed on the basis of time, not distance. Tariffs are equalized within each access level. The Organic Telecommunications Law (LOT) is finally approved, forms the highest juridical framework. Videotex remains a "monopolistic" service, although the European Green Book will eventually force the liberalization of basic-data services, such as Videotex.

During the III National Videotex Congress, Telefónica announced its big Videotex "Promotional Plan" which will be explained further in this paper.

In 1990, the firm "Videotex Información, S.A." was established, in order to provide an impulse to the city of Barcelona where the forthcoming Olympic Games (1992) are to take place. Such an impulse consists of granting financial support and know-how, in creating Service Centers around Barcelona's metropolitan area. In June 1990, the Videotex User Association (AUVE) was founded under the following leitmotif: "Videotex as mass medium, and the domestic user as main actor".

In January, 1991 the levels 032 and 033 were fully operational. July of that year maked the start of "Guiatex", a Service Center guide: a) with the facility of re-routing connection without having to return to the Ibertex Access Center, and b) with more expanded search options: by Service Center number, by firm, by service, by mnemonic, by city, by Autonomous Community and by several criteria simultaneously.

Policy analysis

MAIN ACTORS

The main actors involved are network operator Telefónica, the Videotex Service Providers (APV), including both equipment and information providers, the Videotex User Association (AUVE), and the Ministry of Transport.

Telefónica has lost the monopoly on terminals but has kept control of the network, presumably until 1993. This firm is private de iure although the State is the main share-holder (47%) and no other single share-holder has more than 5% of the total shares. Telefónica is politically controlled by the Ministry of Transport, by the Government as well as by a permanent Government Delegation. The President is elected by the Government while the firm hierarchy is elected from among the people affiliated to the ruling political party (currently the Socialist Party). Deregulation and liberalization are foreseeable in the near future. Yet, Telefónica's political and industrial power remains among the most important in Spain: 11.5 million installed lines, one billion pesetas (7776 MEcus) worth of billing annually, more than 70000 employees, and in 1990 an investment in electronic switching and network equipment exceeding 300000 Mpts (2308 MEcus).

The Videotex Service Providers (APV) is responsible for Videotex development in Spain. Presently it has 75 members of which 32 are information providers, 22 are intermediary information providers, and 21 are software/hardware manufacturers and commercial distributors. The Videotex User Association (AUVE) is the most recent actor whose arrival has met much approval due to its balancing effect between service providers and the information carrier. The Ministry of Transport controls telecommunications (post, telegraph and telephone) through the Telecommunications General Directorate. It, in turn, controls Telefónica through a permanent "Government Delegate" positioned in the company.

GOALS OF THE MAIN ACTORS

In regard to videotex, Telefónica's goal - unlike the French experience - is to create a service, in order that users buy the tool because of its usefulness. It regards Ibertex as the "public service with the greatest growth potential for the next coming years". Among other long term Telefónica's goals, the second-generation, broad-band videotex is worth mentioning. The Videotex Service Providers (APV) aims to defend its members' interests, to promote Ibertex, and to defend the free circulation of electronic information with respect to copyrights and data privacy. The Videotex User Association (AUVE) wish to help the development of a "domestic (home) audience critical mass" which, once translated into demand, will eventually force its own growth. It is relevant to note once again that AUVE believes that Videotex is most suitable for home, not professional users.

The Spanish Ministry of Transport has taken advantage of EEC subsidies, particularly the STAR Program, to develop Videotex in Spain. Subsidies are regulated through a Ministerial Order of May 9, 1988.

The STAR Program, in the area of "Promotion of Supply and Demand of Advanced Tele-communication Services" (article 4.2) has spent 30% of its total budget (18,527 Mpts; 1,425 MEcus) toward such promotion. The subsidization of Videotex terminal is to be considered within the framework of such STAR promotion: 2,250 Mpts (173 MEcus) have been spent during the 1987-1991 period to subsidize some 150.000 Videotex terminals in 80 Service Centers.

Policy however is more interesting than the tables themselves. The STAR Program's Steering Committee made the decision to promote both the supply side (creating Service Centers) and the demand side (subsidizing terminals) of the venture. Accordingly, in order to obtain a grant, the petitioner has to comply to the following requirements: 1) submit a concrete plan regarding the creating of a Videotex Service Center; 2) listing of the expected users; 3) justify the usefulness of the information to be provided and assurance that the information does not overlap with any existing offer of the same kind.

Since the STAR Program is to conclude its activity at the end of 1991, the Spanish Telecom General Directorate plans to use the TELEMATICA Program (1992-1993) to continue subsidizing advanced telecommunication services, including videotex terminals as it did through STAR. How many these will be, and how much money will be involved, will be made public when the EEC divulge the total on grants.

MEANS USED BY THE MAIN ACTORS

Telefónica has three lines of action: 1) technical updating of the "Red Iberpac" (Data Network) - a process which is drawing to a close; 2) promotional, spending money for strategic areas (see paragraph below); and 3) advertising.

In the course of 1990, Telefónica carried out an important "Promotional Plan". It invested 1750 MPts (13.5 MEcus), 750 MPts (5.8 MEcus) of which were used to subsidize 100000 PC communication slots which in turn were sold to the general public for 6000-8000 Pts (46-62 Ecus), 500 MPts (3.8 MEcus) for Data Base promotion, and some 500 MPts (3.8 MEcus) for general Ibertex promotion. Telefónica's plans called for 240000 terminals by the end of 1990, and, as has been mentioned, by June 1991 only 275000 were operational.

The Videotex Information Providers activities are the following: 1) research into the sector; 2) provision of training in the creation of Service Centers; 3) research on Videotex as a system; 4) publication of written information about videotex (Magazine, Guide); 5) participating in Videotex general development policies; 6) international activities; 7) creating of its own Service Center; 8) consultant activities for members.

The Videotex User Association (AUVE) is productive in different sectors: 1) publication of a magazine; 2) creation of its own Service Center; 3) supply of training in Service Center Management (despite the fact that this is clearly the APV's responsibility); and 4) home audience research. The home audience research activities are done not only among real users (in the sense of teaching them new applications) but are also directed toward potential users.

POLICY-ARENA

Recent sociological research shows that the Spanish people are reasonably favorable toward current Information Technologies. The positive values associated with information technologies are well-being, material progress, comfort and efficiency. Negative values are control and dehumanization. Experts have determined however that the lag in the demand for technology is due to commercial inefficiency.

A second general scenario is the Government policy regarding technology. A consensus exists that much more could be done in R & D and in investment in strategic areas. Legislation lag related to Videotex is extensive, and laws become obsolete almost before implementation (this is particularly true in the case of the a for mentioned Organic Telecommunication Law - LOT). Most legislative effort is required in the area of information provision and user privacy.

Telefónica's extensive efforts during the 1990 have led to good results, but it would seem that no more money is about to be made available for videotex promotion in the near future other than for small scale publicity.

The activity of information provision is becoming increasingly disorderly not only due to the lack of a legal framework, but also because of the non-existence of technical standards for the creation and managing Service Centers. The many different protocols which are available do nothing but confuse the users. In addition, the information providers face two structural problems: 1) there is a shortage of specialized professionals in Service Center management; 2) the information providers are caught in a vicious circle, namely that they require home users, since professional users are neither varied nor mobile, but the amount of home users is not growing sufficiently in order to justify investments in the expansion of Service Centers.
Suppliers of both hard and software equipment enter the videotex arena with a varied but not particularly friendly offer. As a result, the average potential user is incapable of understanding the so often cryptic Instruction Manuals as well as the system software.

Eventually, the debate that seems to prevail related to 1) the use of videotex, is it for professionals, semi-professionals, or private individuals?; and 2) where videotex will be used, in the office, or at home?

The Videotex Information Association (APV)'s position is as follows:

- If videotex is for professional <u>computer</u> use, the answer is absolutely no. In this respect, videotex (hard + soft + network) is an inadequate system.

- Videotex is geared toward personal, non-professional use in the home.

Neither assumption is based on the current ... The situation may change 1) when services become truly relevant to individual users, 2) when tariffs are lowered; and 3) when the number of users increases whereby the quantitative and qualitative growth of relevant information will be made possible as well as the lowering of the price for Videotex Service.

Telefónica's point of view is that the dual orientation is not valid because the service can be of use to both types of users, despite the fact that videotex was originally geared towards private individual users. In countries like Spain, Ibertex must first expand within the professional market, and from there widen its scope to include the non-professional users.

Finally, it is obvious that the Videotex User Association (AUVE) strongly supports the idea that Videotex is a non-professional, bi-directional mass-medium.

WHO PLAYED THE DOMINANT ROLE IN THE VARIOUS PHASES

Up to two or three years ago, it was obvious that all industrial policy related not only to telecommunications but, to a lesser extent, to electronics, was conceived, managed and controlled with a will of iron by Telefónica. Nowadays, taking into account the foreseen liberalization, this organization has rid itself almost completely of this control and no actor has emerged to play a substantially dominant role. It is worth mentioning however, that Telefónica, as a commercial firm is acting more and more like one, and is very much interested in the commercial success of Ibertex since traffic is an intrinsic part of that service. However, Telefónica refused to invest the hugh amounts of money that it did in 1990 to promote Ibertex. That means quite plainly that it has no intention of remaining a dominant actor in the videotex arena.

THE MOST DRAMATIC SHIFTS

Given the **de facto** public character of Telefónica, it is not surprising that the two most important moments of growth for Ibertex are related to two political events: the general elections of 1986 and the World Football Championship of 1982.

However, it is obvious that the most dramatic shift was due to the Telefónica grants in 1990 as well as the STAR subsidies which were directed towards videotex within the general program of promoting advanced telecommunication services in the less developed areas of Spain.

According to the Videotex Service Provider Association (APV), the most recent, most dramatic and positive change has to do with the establishment of the kiosk system. This development has illustrated that there is business in providing information. In April 1991, the information providers started receiving money. So far, although no reliable data are available, all "signs" suggest that things are moving ahead. The kiosk system centralizes quality information, since there is money to be made by Service Centers which offer quality. It also introduces a competitiveness because information supply has entered into a free and competitive market.

A third important shift should be mentioned. One which is related to the general debate "private/public network". The star of this debate is the "Banco de Santander".

"Banco en Casa" (Home Banking) is a trademark of Banco de Santander. This term comprises three telematic realities: telex, telephone and videotex home banking. All three are still operational. The Telex Home Banking, initiated in 1985, uses the telex network which is managed by the Spanish Post and Telegraph Office, since Telefónica has become a private company. The Phone Home Banking (quite similar to the Audiotex) uses a small, pocket-size dial and acoustic modem which can be connected (that is, fitted) to a regular phone terminal. Data transfer from the server to the user is done by voice synthesis. Date transfer from the user to the server is effected through the multi-frequency dial.

The first "TV Home Banking", as it is called, was initiated in 1985 following the CEPT-2 norm since Minitel was considered the model to be followed. The Banco de Santander was granted authorization by Telefónica to create and use a private network. Obviously this agreement broke the legal monopoly. It seems reasonable to think that Telefónica's loose interpretation of the law and broad-minded cession of its legal rights were permitted as a sort of testing ground of the forthcoming liberalization of the Telecommunication market, not to mention the pressure exerted by the Bank.

By 1990, the Banco de Santander initiated the standard Videotex Home Banking using the public X-25 "Iberpac" network. Nowadays, both public and private network systems are used, although the trend and the policy is to increasingly shelve the private network adventure. Why? Two main reasons have been given by the Bank spokesman: 1) The cost of creation and maintenance of the private network is higher than the renting of the public one; and 2) from a commercial point of view, the private Videotex network is a closed user system (encompassing just the Bank's clients) and as such it was becoming more and more isolated from the open, public network Videotex system both at the national and international levels whose access is open to all Videotex users.

SPECIFIC ROLES OF THE ACTORS INVOLVED

The complete set of actors and their relation to one another is clearly illustrated in figure 1.

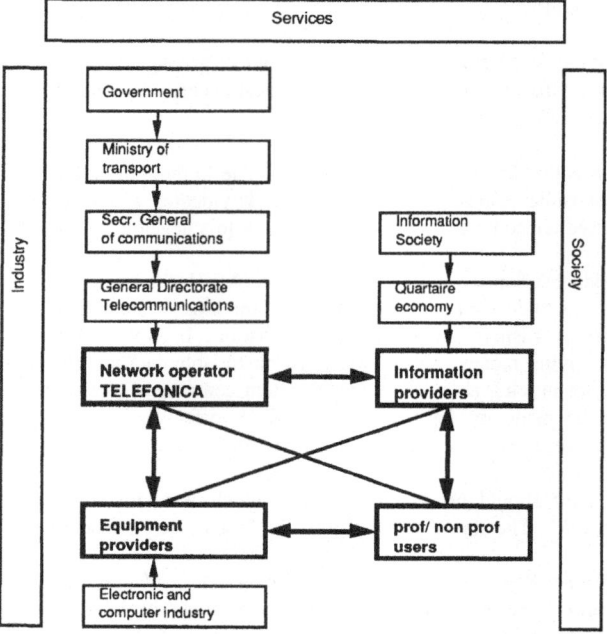

Figure 1: Four main actors and their associates

The main actors are: 1) the network operator; 2) the equipment providers; 3) the information providers; and 4) the users. The first two are located in industry, namely telecommunications, electronics and computers. The network operator is strongly tied to the political environments, both national and European. The network operator and information providers are part of the services current trend toward informationalization, the base of a growing part of the economy. The information providers and the users are situated within the broad societal arena. The first three actors offer the service, the fourth requests the service.

As in any other human realms, each actor plays his role, defends his own ideas and criticizes the ideas of others. But all too often, his behaviour is based upon hypothesis, not on facts. We argue that a comparison of this theoretical model is needed at the empirical level in order to analyze the dynamics of the variable interplay involved.

Conclusion - specific characteristics or problems of spanish videotex

Spanish Ibertex, one of the most important European videotex behind the French "giant", has several specific characteristics:

    1)   It has an important User Association
    2)   The Providers Association is composed of both equipment (hard + software) and information providers. This situation, generates conflicts and will soon

have to be solved by splitting responsibilities.

3)   Telefónica, the Spanish Network Operator, is **de jure** a private company but **de facto** a public institution still heavily dependant on the Government.

4)   Despite the fact that a 7% of Spanish homes have a PC, Videotex apparently is most popular by professional users while private use takes place outside the home. Fear of tariffs?

5)   The Ministry of Transport has channelled important STAR subsidies to promote service centers and has given grants for buying terminals and slots.

6)   A private bank (the "Banco de Santander") initiated a private network vtx service which has remained unsuccessful.

Furthermore, there are two main problems:

1)   The forthcoming Telefónica's divestiture and the liberalization of the telecommunications industry have created a rather confusing situation in the videotex arena.

2)   The general inability of persons and institutions to enter the risky business of Service Center managerial activities slows down the establishment of Service Centers.

Future perspectives

The future of Videotex depends on the interplay among the four above-mentioned actors and their associates. Also we suggest that the future' of Videotex will strongly relate to the following trends:

1)   Network: the quality standards must be further improved; broad band network is imperative for the future success of Videotex because it will favour interactivity as well as dynamic video information.

2)   Equipment: whether dedicated or on PC, the question is directed toward software user-friendliness and conviviality.

3)   Information: In the midst of a current information explosion, the information society has to decide at long last what type of information is truly relevant to individuals. This appears to be the real challenge of Videotex.

4)   Users: It remains to be seen whether videotex is a system oriented toward professionals or non-professionals. Statistics provided by European operators regarding the number of calls and connection hours purposely seem to avoid distinction between these two types of users.

CHAPTER 10

SWEDEN: THE TROIKA PATTERN

Helena Flam
Universität Konstanz
Socialwissenschaftlische Fakultät
B.R.D.

Joanna Rose
University of Lund
Institute of Sociology
Sweden

Introduction[67]

Despite some technical and administrative shortcomings, Swedish videotex has to be seen as a relative success insofar as it managed to attract more subscribers than the Austrian and Danish videotex systems. In per capita terms, Swedish videotex seems successful even in comparison with FRG and Great Britain.[68]

This "relative success" is surprising when one considers that Swedish videotex has never been defined as a spearhead of a communication revolution. In a striking contrast to Germany, Great Britain and France, Swedish videotex received no government-support. Neither Swedish interest organizations nor Swedish political parties took any special interest in its development. Instead, Swedish videotex has been a purely commercial activity. Its major public developer - Televerket - has never treated it as its top investment priority.

The "relative success" of Swedish videotex is also surprising in view of the technical and administrative obstacles it encountered within Televerket where the developers of videotex within Televerket have suffered from status deficiency. Between 1975 and 1989 they had to negotiate important decisions with the traditionally strong net- and telephone-divisions. Moreover, the development of videotex network in the country had to be negotiated with 20 separate regional division heads. First in 1987 the videotex "experiment" finally acquired a status of a separate and independent project with its own project director, directly responsible to the director of Televerket.[69]
Finally, the "relative success" of Swedish videotex surprises when one considers that its information providers (IPs) have criticized its technical features and ineffectiveness on several occasions. Many large firms, including major banks, chose to start their own systems or to leave

---

[67] Our research has been supported by the Bank of Sweden Tercennary Foundation. We would like to thank videotex developers in Sweden, including those at AU-System, Telebild, Televerket and the Swedish Videotex Association, who kindly provided us with the information on which this chapter builds. For reasons of confidentiality their contributions remain anonymous. However, it must be noted that the report is based on about 70 expert interviews. A more complete research report, entitled "Introduction of Videotex in Sweden", may be ordered at Sociologiska Institutionen, Lunds Universitet, Box 114, 221 00 Lund, Sweden.

[68] There are 270 citizens in Sweden per one videotex subscription, while in Denmark there are 3571, and in Austria 840 citizens. FRG has one videotex subscription for 635 German citizens while in UK there are 700 citizens per a subscription (see Flam and Rose, 1990:14).

[69] From July 1991 videotex became incorporated and is now part of Telemedia Inc., a subsidiary of Televerket.

H. Bouwman and M. Christoffersen (eds.), Relaunching Videotex, 125–140.
© 1992 *Kluwer Academic Publishers.*

and ignore videotex altogether. Moreover, at least a fifth of videotex subscribers never uses videotex services (Fa 47/89: 1,5).[70] About 40% of the information users (IU's) have expressed dissatisfaction with the videotex system (Fa 6/91:4).

For some time now, Swedish videotex-developers have realized that the videotex market is stagnating. This paper will attempt to elucidate why this happened, the main thesis being that the orientation of the Swedish videotex developers to business accounts in part for both its early successes and its current stagnation.

Short History of the Swedish Videotex System

In the 1970s, Televerket - a Swedish administrative body with an informal monopoly on telecommunications - started preparations for its own upcoming privatization and the deregulation of the entire telecommunications sector. These were expected to occur in the 1990s. Televerket's market division planned and looked for new strategic areas of development. Among them were videotex, data communication as well as telex and telefax. Investments in videotex were not large in comparison with the other new technologies (Teldok 1987:290). Videotex seemed particularly easy to develop. Sweden had the densest, most frequently used, and cheapest telephone network in the world (Teldok 1987:111-120). It also had a highest number of computer connections per 1 000 employees in Europe (Visionen 1985:1).

The efforts to develop a Swedish videotex technology paralleled the British from 1975 on. The early Swedish version of videotex built on the Prestel concept. The developers of its Swedish equivalent followed its progress closely. The first ambitious Swedish videotex pilot project - "Datavision" - started as early as 1979, a year after the British field trial. Moreover, the first two videotex data bases and videotex systems (Televerket and Philips) appeared in the late 1970s. They aimed at developing methods for page storing in computers connected by telephones to user terminals (adapted TV-sets). The system architecture, based on the Prestel model, was hierachical and the videotex information, as in Prestel, was stored in a tree-formed structure. The first videotex system consisted of a computer, a memory disc and eight access points. The decision to follow the British Prestel concept caused many technical and market problems which slowly led to its redefinition.

Televerket was mainly occupied with the technical aspects of videotex. However, on the information-providing side, Televerket cooperated closely with the "Telebild"-project from the moment it started in 1980. Svenska Dagbladet (a big daily), the Central Swedish Employer Association, the union Industriförbundet and SE-Banken co-sponsored Telebild (Ohlin 1986:81, Teldok 1986). Thirty-two big and small enterprises, among others Ericsson, Volvo and SAS, participated in the Telebild project in 1981/1982. Rather than starting their own videotex systems or hiring space in a computer, the project participants decided to cooperate with Televerket to draw advantages from its already accumulated technical experiences. Many of these participants later started their own systems or decided not to use videotex at all.

Datavision became commercial in the spring 1982 and simultaneously, Telebild launched its

---

[70] Televerket's marketing division conducted phone surveys in 1986, 1987, 1989 and 1990 . The random sample included 210 persons in 1986, while it included 500 persons in 1987. In 1989 the survey included 454 persons and in 1990 it included 500 persons (Rapport Fa 53/86; Rapport Fa 41/87; Rapport Fa 47/89: Rapport Fa 6/91). In the 1989 sample, however, only those actually using videotex (353) were asked to answer all questions. Televerket kindly provided us with the summaries of these surveys.

services as a private firm - Svenska Telebild AB. Together Telebild, Televerket and AU-System, a software firm, formed the Swedish "troika". The decisions of this "videotex troika" shaped and exerted long-lasting influence on the development of the Swedish videotex.

In contrast to the technological videotex concept, Sweden did not repeat the British mistake of choosing households as the most likely videotex users. If by anything, the "relative success" of the Swedish developers can be explained by this decision of Televerket and Telebild. In particular, they focused on identifying business "niches" in which there was a need to rationalize information retrieval or flow. The second mark of the early marketing strategy was "information-packaging". AU-System worked out the technical aspect of "information packaging" for each client group, including both IPs and IUs. These very first decisions still affects the functioning of the videotex market today. It accounts for the predominance of business-oriented information and of professional users.

Already in 1982/83 the Swedish videotex concept underwent a change. It no longer stood for pages stored in a central computer, but for data transmission between one central and several external computers. However, the new concept could not be put in practice as long as the gateway problem remained unsolved. In Sweden, videotex developers did not underestimate the scope of this problem the way their German counterparts did (Schneider et al. 1990 7). They started to look for a solution from the beginning of the commercial phase and not in the midst of it as a means of alleviating a negative market response to videotex. The solution, a Swedish invention, was a transaction-oriented, permanent connection to the videotex system. It allowed external computers to be connected to Datavision more than two years before Prestel gateways appeared on the Swedish market at the end of 1986. Since then Televerket's Datavision had a privileged status as a main IP, these external computers had to transmit information through it. The mediation of Datavision could not be circumvented at this stage. Therefore, the Swedish system, although no longer completely centralized, still differed from the decentralized French system.

Three factors slowed down both the technical development of videotex and the marketing efforts:
a)   Televerket's multiple roles as both the network provider and the owner of the biggest database in videotex
b)   the marginal status of the videotex project within Televerket
c)   the prolonged focus on the technical development and administrative management.

The focus on technical development meant that no separate marketing personnel was included in the videotex project and that no sustained marketing attempts were made. The preference for sophisticated standard technical applications prolonged the necessary research time. The marginal status went along with relatively limited resources. These two factors delayed the commercial phase of videotex untill 1982. Finally, in an attempt to retain its role as an owner of the biggest data base in videotex, Televerket made communication between external computers and IUs rather costly. IUs who had to go through Datavision to a desired data base, paid separately for the "second connection" and for using lines between Datavision and external computers. The additional fees lessened the attractivness of videotex among businesses with or in need of access to external computers.
Televerket made an explicit effort to expand the videotex market in 1985. It announced its intention to invest 300 000 000 Skr (375 000 ECU). Several months later, it decided to import the IBM/BTX system. Both these gestures were meant to encourage large-scale, professional IPs to join videotex and to pull the IUs with them. However, Televerket did not achieeve this goal. On the contrary, the technical development produced many problems, especially of access, speed and overall system effectiveness. Televerket also changed administrative routines and prices. These problem-causing changes slowed down the expansion of the videotex market.

The decision to buy the German videotex system and its implementation in 1988 caused two big waves of protest among IPs and IUs. The decision was taken without any consultation or negotiations otherwise so typical in Sweden. Moreover, IBM/BTX was hierarchical and centralized. Videotex users preferred the decentralized French Minitel. IBM/BTX required IPs and IUs to make changes in their old systems. It furthermore turned out that the EHKP gateway protocol for the external computers was expensive and time-consuming. Swedish businesses were not really interested in changing the presentation standard from the relatively simple Prestel to the more advanced, but also more expensive CEPT.

In 1979-80 Televerket expected videotex to have 100 000 subscribers in 1985 (OECD 1988:15). Seven years later, it expected 80 000 subscribers by 1990. Inflated at first, Televerket's predictions were deflated by reality. Televerket's own 1989- and 1990-surveys[71] showed that more than two-thirds of all firms using videotex did not intend to purchase more terminals or subscriptions. Finding new business niches for videotex seemed increasingly problematic. In the late 1980s, all videotex developers spoke about the "saturation" of the videotex market. The reason for this was that the professionals were not interested in the simplistic, inneffective and expensive videotex technology. The decision to switch to the IBM/BTX apparently failed as a strategy for attracting professionals.

Our research indicates that the opinion of the professionals should not be seen as the sole reason for the "saturation" problem. The decisions taken by the "videotex troika" contributed to this problem. First, information for business hardly attracts household users. Their share among IUs consequently fell from about 10 to about 1 per cent. Secondly, the videotex system has been ridden by serious technical problems. It has been ineffective as a consequence of the early decisions of a technical nature. Hundreds of videotex users not only protested but also gave up their subscriptions during the last decade. Yet, Televerket has remained unresponsive to the needs of its own clientele. It does not seem to need the advice of its own marketing division. This division twice recommended that the access, speed and information-finding ought to be improved before new marketing attempts were undertaken.

The current state of the videotex market can be characterized as that of consolidation and intensification rather than expansion. The way out of the saturation problem is sought in a new videotex project: **TeleGuide**. It aims at enlarging the market by creating a parallel, household--oriented videotex service, addressing publics with well-defined needs and consumption patterns.

This project is inspired by the French Télétel system. In the beginning of 1988, Teleguide distributed 50 Philips' Minitels and 50 Finnish Salora terminals free of charge in Västerås. Televerket (although not fully engaged) and 15 big Swedish enterprises (including banks, mass media, insurance and travel companies and post order firms) supported this experiment. The plan was to start the second phase of the project by spreading 50 000 videotex terminals in January 1989 and to establish a corporation.

In January 1989, IBM, Esselte and Televerket took over the TeleGuide initiative. This new "videotex troika" combines the three basic videotex ingredients: network infrastructure (Televerket), videotex technology (IBM) and information (Esselte). Therefore TeleGuide is in a position to decide on how, for whom and with whom the system is going to work. Televerket's participation can be seen as gesture testifying its good-will. The TeleGuide investors intend to spend about 250 millions Skr. Banks, travel- and insurance companies, mass media and post order firms are, as in the beginning of TeleGuide, involved in the project, but this time as IPs.

---

[71] See note 3.

The history of TeleGuide brings us back to the importance of the troika pattern in the history of the Swedish videotex. This pattern became visible already in the experimental and early commercial phases. TeleGuide's idea of packaging videotex information resembles Telebild's policy except that this time the target is a private user.[72]

But in 1991, three years after the new troika took over TeleGuide, the project still waits for a green light.[73] Although a small organizational network has finally been stabilized and in-corporated itself as a TeleGuide company, the prolongued discussions about the choice of technology, billing systems and the type of taget users, etc. are still going on. One important matter under discussion is the connection between the "old" videotex system and TeleGuide.[74] The "old" videotex IPs and IUs have already gained some advantages from the discussions inside TeleGuide: time tariff, requested for a long time, has been introduced in the Televerket's system in the spring of 1991. When TeleGuide will start, the faster communication (2400 bps instead of 1200 bps) will become available for all IUs in the system. Some decisions of the technical nature have already been made. For example, the decision to buy 50 000 German-made Loewe terminals. The plan is to spread a total of 5 millions terminals in the next five years. This terminal type is already in use in Switzerland. It can work with a "smart card". The card handles user-identification, purchases and financial transactions. It is simple to use and is said to solve the security problem. TeleGuide expects that, encouraged by the "smart card", such IPs as travel, insurance, post order, and, most importantly, banks are more likely to join TeleGuide and make it even more attractive for private videotex users.

Even if the idea of TeleGuide services is French, TeleGuide seems technically more advanced, with its Loewe terminals and "smart card", than the French Minitel. As a "late comer" on the home market, Sweden learned from both German and French lessons.

General Description of the Current Videotex Situation

INFORMATION PROVIDERS AND USERS, THE OFFERED SERVICES

The major videotex developers in Sweden oriented themselves from the very beginning almost exclusively towards business. Among the main users of videotex are smaller firms dealing with commerce, banking and insurance, car- and real estate-dealers, etc. There are about 100 IPs and about 55 external computers. Their number is expected to grow when the TeleGuide project starts. There is a general tendency of concentration of service provision - Esselte already bought Telebild in 1987 now buys other service providers.

---

[72] This is no coincidence since the director of Telebild gave birth to the TeleGuide idea and was hired by the new troika as a consultant.

[73] The plans of TeleGuide are still not quite clear - it has already several times announced the start. The latest news is that 70 000 terminals will be spread before April 1992, although only about 10 000 will reach the IUs before January 1992, as the Teleguide plans for a "soft" start (Interview, September 1991).

[74] The information-providing part of the TeleGuide company, called TeleGuide Scandinavia Inc., is owned in 90% by Esselte and in 10% by IBM. It is to be treated as an external IP with an own subscriber group in the public videotex network (an "umbrella" organization for TeleGuide's IPs). Therefore it has a right to decide about a fee for entering the public videotex system through TeleGuide. TeleGuide decided a price of 15 Skr (2 ECU) for one call. This has become an area of conflict between Teleguide and the "old" IPs and IUs, who protested against this fee.

Table 2. Swedish Videotex Subscribers

Year	Number of subscribers	Traffic minutes
1981	170	
1982	300	
1983	1 823	
1984	4 536	
1985	7 000	
1986	9 875	
1987	13 300	35 600 000
1988	16 683	45 900 000
1989	23 264	52 700 000
1990	27 000	about 48 000 000[12]
1991	30 000	about 48 000 000

Source: Televerket, interview, September 1991.

Televerket's videotex system has about 30 000 subscribers in 1991. The growth has not been as smooth as Table 2 suggests. There were periods of stagnation in 1985 and 1988 when Televerket was making technical and administrative changes in the system.

The most striking impression from Televerket's 1989-survey[75] of videotex users is that they are evenly divided among those who find that videotex fulfilled their expectations and those who are disappointed. It is a notable phenomenon that 50% of the videotex users are dissatisfied, that only 41% use it daily, and that a fifth of subscribers never uses videotex services.

Televerket's user surveys from the 1980s show that a majority of the users are men, working for firms with less than 20 employees. Small firms clearly constitute the current base of the Swedish videotex market[76]. The most frequently represented business is commerce. Between 1989 and 1990 banking and insurance ranked second. By 1990 services and consulting became the second most frequent videotex user.

Among the most used services is the telephone directory. Until 1989 it was followed by financial information, "special-package" information for car dealers and real estate market, and travel information. In 1990 travel information has become the second most used service. As in 1989 so in 1990 only 40% of IUs use videotex daily.

TECHNICAL SOLUTIONS

During the experimental phase Datavision had only one computer, but in 1982 one more computer was put into operation. The number of access points was increased to 64. However, even if the number of users was not large, it was hard to get a connection because it depended on the number of callers and the connect time. In 1982/83 it was impossible to make an direct connection to either Stockholm or Gothenburg. In response to criticism, but also because of optimistic expectations, Televerket installed a 4-computer system with 192 access points in 1984. It also made plans for a regionalization of videotex in 1986 which were forsaken when the

---

[75] See note 3.

[76] In 1990 firms with no more than 5 employees constituted 41% of the sample (Fa 6/91:1).

decision to buy IBM/BTX was taken. The quality of the Swedish telephone network also varied and caused troubles with access to the system. To overcome the transmission problems, new multiplexors were bought and gradually installed in the telephone network during 1986. They did not, however, suffice as an improvement.

The second problem was the speed of communication. The first generation computer could use only the 75/1200 bps velocity. Since 1982 the IPs could use some access points with 1200/1200 bps full duplex but one still had to wait up to 3 minutes before a page was in the system. In 1984 AU-System installed permanent connections between external computers and Datavision with velocities of 2400, 4800 and 9600 bps. The same year, users could also call Datavision through Datex, the circuit-switched data communication network. These new data transmission methods were faster but more expensive and until today very few videotex subscribers rely on them.

The third problem concerned connecting external data bases to Datavision. AU-System developed the first gateways in 1984 and the first standardized gateway solution came two years later. It was a Swedish adaptation of the British Prestel gateway coupling - the PEGY protocol. A year later, EHKP, a German standardized connection which could work both the CEPT and Prestel terminal standards were released.

Terminals constituted the fourth problem area. In 1980/81, when business became defined as a market target, Telebild and Teledatorer started looking for "professional" terminals more suitable for business than the adapted TV-sets. At first they imported Austrian MUPIDS. These proved to be a "technical disaster". In 1982, in a parallel effort, Telebild placed orders with a small electronic firm - Registrator. Registrator built VX 150 and VX 200 which became the most popular terminals in Sweden by 1986. They had automatic functions, including automatic log-in. Salora, a Finnish producer, was Teledatorer's main competitor for a while. Together Salora and Teledatorer had large shares of the Swedish terminal market. They faced no real competition until the PCs appeared in Sweden in large numbers in about 1987. In 1987, already 40% of non-frequent and 13% of frequent IUs worked with PCs. Teledatorer, seeing this trend, switched to PC distribution and software development for videotex. By 1989, PCs clearly dominated among all types of videotex users, and in 1990 62% of Swedish IUs had a PC as a videotex terminal.

Log-in and search procedures are still found very problematic among videotex users. Both in 1990 and 1989 search through page numbers was the least preferred information finding method. In 1989 24% of IUs thought that it was very difficult to find information in videotex. An automatic log-in, "quick-choice", pre-programmed and personal menue terminals seem to have been the best working technical solutions, adopted by the "most frequent" users. In contrast, manual log-in and the search procedures such as subject matter catalogues and search words bred higher rates of dissatisfaction.

The way out of the technical impasse caused by the inefficient videotex network was, as Televerket saw it, to change the system in Sweden completely. Televerket decided to buy the German IBM/BTX which became operational in 1988. Technical changes and the changes in administrative routines accompanied the new system. Some external computers still have problems with connections to the new system. Since the old gateway protocols were forbidden by Televerket in 1988, the external computers had to connect to the IBM/BTX system through Datapak and the EHKP gateway protocol. These turned out to be complicated and unnecessary expensive.

The new IBM system is considered too slow for both IPs and IUs. It is also expensive and

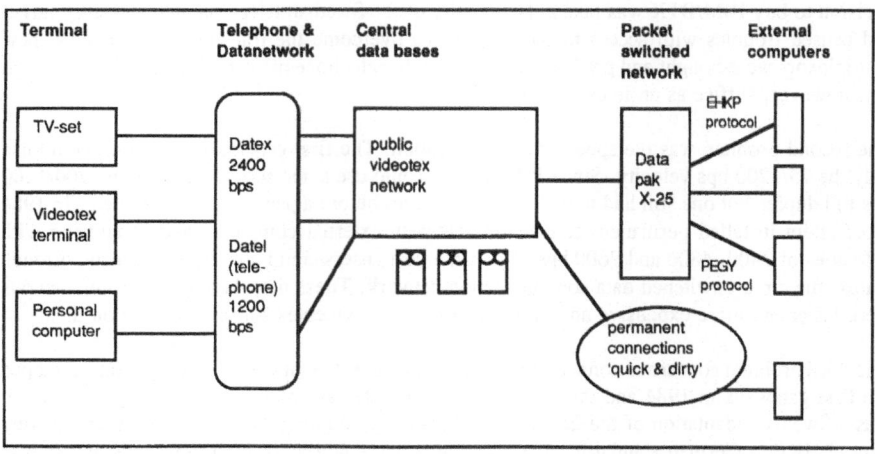

Figure 1: System Architecture of Swedish Videotex

cumbersome for IPs. In 1991 Televerket opened a few 2400 bps connections for CEPT users. It promised to open more later on, when the TeleGuide project starts. TeleGuide is to work in the CEPT standard, although it is not widespread.[77]

BILLING SYSTEM

Charges for information users are shown in table 1.

Table 1: User tariffs in the Swedish videotex system

Installation fee	250 Skr (32 ECU)
Subscription per month	30 Skr (3.75 ECU)
Extra subscriber per day	0.65 Skr (0.08 ECU)
Connection fees:	
min. in peak hours	0.73 Skr (0.10 ECU)
min. in off-peak hours	0.56 Skr (0.07 ECU)
Sending a message	0.50 Skr (0.07 ECU)
Storage per message per day	0.10 Skr (0.01 ECU)

The users are billed for each page they request to see. Information providers pay for the toll-free pages. Page tariff can vary from 0.01 SKR (0.001 ECU) to 99.99 Skr (12.5 ECU) per page.

Different services have also extra subscriptions and/or extra page charges. For example, Televerket charges IPs for "closed user group" service, while every IP decides about subscription prices.

---

[77] In the 1990-survey, 33% of IUs had Prestel graphics and 21% had CEPT. 46% did not know what videotex graphics they had (Fa 6/91:4).

As regards the charges for information providers every IP pays 5% of his income to Televerket for the costs of the system administration. Televerket administers all payments from IUs to IPs.

The connection of an external computer to the videotex system costs 25 000 Skr (3 125 ECU). An extra computer for the same IP costs 15 000 Skr (1 875 ECU).

Policy Analysis

MAIN ACTORS

Swedish financial firms, some industrial interest organizations, such as SAF and Industriförbundet, and one national daily took the initiative, along with the state-owned Televerket, to test and develop videotex in Sweden. After a test period, the Telebild project was established. Together Telebild, Televerket and AU-System, a software firm, formed the Swedish "videotex troika". The troika implied a (rather hazy) division of labor between the three main actors. It involved and combined three distinct areas of responsibility for videotex: system network and administration (Televerket), services (Telebild), software (AU-System).

Until 1987 Televerket was both an owner of the network and of the main data base in the system. Datavision's staff worked both on the technical and the market (IP-recruitment) side. Telebild was involved in terminal sales and IP-recruitment, while AU-System worked with technology, but also dealt with terminals and IP-recruitment.

A distinctive feature of the Swedish videotex development has been the very small size of the central videotex developing group. Both in terms of "interlocking directorates" and in terms of the number of the persons involved, we can speak of a tiny elite as the dominant developer of videotex in Sweden (Mills 1961; Domhoff 1979). As far as interlocking directorates are concerned, Televerket was co-owner of AU-System and became a co-owner of Telebild, when it became incorporated in 1982. Among the first owners of Teledatorer, a major supplier of videotex terminals in Sweden, were AU-System, and Registrator.

Apart from "interlocking directorates", one can also speak of "interlocking staffs". Teledatorer's and Telebild's staffs were almost identical in the beginning. The offices of these two firms were located on the same floor, and staff members only needed "to change hats" to work for one or the other. Moreover, AU- System's staff often performed tasks directly for Televerket.

In effect, a very small number of persons and institutions have been making decisions about Swedish videotex. They have changed its concept. They have made successive choices of the target users, system architecture, terminals, appriopriate market strategies, etc.

The members of the first "videotex troika" managed to exclude competitors, such as, for example, Philips and Salora. They established a domain within which they set up operating rules. Within this domain Televerket monopolized slow modem provision until 1983 and for long remained a major data base/IP. Until 1989 it had a monopoly on fast modem provision. Its role as a monopolistic price-setter, which on occasion gives in to outside pressures, has not changed. Similarly, within this domain, AU-System monopolized software provision, while Telebild had a majority of IPs under its organizational umbrella. In 1987 it had 60% of all videotex traffic. AU-System, Telebild and Teledatorer, as the co-owners of Registrator, managed to gain control over the market on dedicated terminals.

This domain became restructured as new actors penetrated it and outside criticism pressured for change. Televerket's role as a major database owner was critized by other service providers. As a reaction to this criticism, consonant with the privatization trend, Televerket sold the service part to Teleinvest, a Televerket-owned financial company, in 1987. Televerket became only responsible for network provision. In 1991 the videotex project became incorporated as Videotex Inc., but it still has financial ties to Televerket. It is now owned by Telemedia Inc., a subsidiary of Televerket. In its turn, AU-System lost its monopoly on the provision of software and instead became an IBM agent, when Televerket's top management decided to let IBM - a new actor - to enter the videotex domain.

On the terminal side, AU-System bought Teledatorer and promoted PCs on the Swedish market. AU-System lost some of its privileged position in the videotex market, it made an overall profit and moved to other markets. From a small firm in 1979, AU-System became the largest software producer in Scandinavia by 1990. Esselte, a multi-product firm, acquired a majority control over the third dominant actor within the videotex domain, Telebild, in 1987. During the 1980s Esselte developed a greater interest in videotex, and ultimately swallowed Telebild. By 1987 as new actors and products entered the stage, the members of the first videotex troika either disappeared or lost some of their privileges.

Nevertheless, the "videotex troika" patterncan still be found in Sweden until today, although the troika has a different composition. Today's troika, in charge of TeleGuide, includes IBM and Esselte in addition to Televerket. Although several attempts have been made to go beyond the troika-structure and to involve a broader array of commercial actors in the videotex developing activities, these have failed.

Swedish IPs, both the big and the small ones, are organized in VIS - the Swedish Videotex Association. VIS was established in 1983 primarily to promote and coordinate videotex but also as a counterweight to Televerket. VIS organizes most of IPs and provides Televerket with information about the videotex market. It has a rather weak bargainig position in relationship to Televerket. Televerket is also a member of VIS.

ACTOR GOALS

Televerket, AU-System, Telebild-participants and even the larger banks which dropped out shared an interest in securing technical competence in a new area of technology. The first three, along with VIS, had also a goal to create a broad market on videotex in Sweden.

Several large banks initially interested in the concept withdrew already in 1982 when the commercial phase of Datavision began. Others closed their public services when they saw that the security problem was not solved and that their goal of developing a national videotex-bank system would not be realized.

Televerket's interests have been in identifying strategic investment areas, but also in defending its privileges - both its monopoly on modem provision and its status as a central data base. As the regionalization plans developed for videotex in 1985, its staff had no particular stake in a strictly centralized system. Televerket's technical staff had as a goal the development of sophisticated and standardized technical applications. This goal put it on occasion in conflict with AU-System and Telebild. These two firms supported "quick and dirty" solutions (expression of Televerket's technical staff).

Telebild strove to become the largest umbrella organization for IPs. AU-System, in its turn, worked to monopolize the software and terminal market. With their clients, they shared a goal of improving and rationalizing the intraorganizational and sectoral communication. Users had a prevailing interest in a simple, quick and cheap-in-use videotex system. They were not very sensitive to the equipment costs. However, they showed themselves very concerned with particular fees and Televerket's privileges and monopolies. In particular they have opposed per-page charges and extra fees for access to external computers.

VIS long-term goal is to promote videotex and to keep it outside the censoring control of the state. It also has a stake in making users to conform to the ethical rules it had worked out. It is VIS' ambition to become a discussion partner of Televerket.

ACTOR MEANS

Datavision relied on the financial resources provided by Televerket and had encountered few problems in securing these resources for the purpose of experimenting with videotex.

The small videotex group within Televerket was dependent in its price-setting of the costs of transmission in the videotex network on the telephone- and net-divisions which were responsible for Datel, Datex, Datapak, etc. In relationship to the videotex users, the price-setting was unilateral, although on a couple of occasions Televerket's top management responded to users' pressures.

Televerket cooperated with a number of firms and played a "divide and rule" strategy in the experimental phase before it tightened its cooperation with Telebild and AU-System in the commercial phase. Datavision attempted to mantain its multiple roles backed by the traditional, institutional power of Televerket and its own pricing decisions. The "interlocking directorates" with Telebild and AU-System insured Televerket of a measure of control over their decisions, in particular those concerning the choices of technology. However, on a few occasions, these two engaged in conflicts with Televerket.

Close cooperation between AU-System and Telebild gave them a certain negotiating power in relationship to Televerket. In turn, their own cooperation, as well as that with Televerket, helped them to increase gradually their importance. Moreover, AU-System has all along worked with other Swedish firms that had their own videotex systems. Telebild, seconded by AU-System, identified niches in need of videotex and made top-level sectoral agreements about their introduction. Both strived to gain greater control of the market development in Sweden.

Swedish videotex users have expressed their protests and complaints in letters to Televerket and in personal conversations with the members of the videotex troika. On a couple of occasions, hundreds of videotex users have given up their subscriptions. IPs, in particular, voiced their dissatisfaction through VIS or in Visionen - a videotex magazine owned by Televerket. They also wrote to the Swedish industrial "ombudsman". Telebild and AU-System often sided with the clients on such issues as Televerket's technology choices, its monopoly on modems and its role as a privileged data host.

VIS formulated ethical principles for videotex which so far have exempted it from excessive government control and regulation. VIS monitors and, if necessary, applies sanctions against deviant IPs and IUs. It also tries to defend their interests against the press when incidents occur. In order to promote videotex in Sweden, VIS has organized various informational activities and participates in international organizations. Moreover, it has attempted to act as a pressure group in relationship to Televerket by writing letters and organizing meetings with Televerket's representatives. It has tried to gain some influence over Televerket's decisions concerning prices and choices of technology.

Between 1977 and 1984 four different investigative commissions concerned with the new medias debated the role and nature of videotex (Ohlin 1986: 136-137; SOU 1981). The work in and outside these commissions has laid foundations for the present-day praxis.

The press, worried by the competition of the new media, called for the first investigations. However, when it noticed in the early 1980s that videotex was business-oriented, it realized that at least videotex was no competitor for mass advertising. Apart from the press, the members of the Social Democratic and Center Parties were pushing for an investigation of the societal and economic aspects of the new media. They became instrumental in the shaping of the agenda of the inquiry. They were concerned with advertising in the new mass media. They also feared that if not controlled by the state, videotex could be used for unethical purposes, such as porno-advertising. In the "Investigation of Information Technology" (1978-1981), a Social Democratic majority, which defined videotex as a mass medium, since like TV it used a TV-screen and involved wire-transmission, demanded that advertising should be prohibited in videotex. A minority, among whose members were the early videotex developers opted instead for allowing "demanded advertising", a special marking of pages with ads.

VIS quickly established a committee devoted to videotex ethics. This committee recommended that advertising should be "demanded" - the user should be informed in advance that advertising is placed on certain pages. By adopting the principle of "demanded advertising", initially suggested by the "Investigation of Information Technnology", VIS managed to defend the videotex sector against any further discussions about forbidding advertising.

The "Mass Media Committee" proposed that videotex should be free from external regulation. It should be subject to the principle of free but responsible establishment and bound by damage and criminal laws. The videotex sector, in its view, should assume responsibility for self-regulation and, consequently, self-police itself. VIS' ethical rules were percieved as a sufficient guarantee that advertising in videotex would not be left unrestrained and unregulated.

No other significant debates about the role and rules pertaining to videotex took place within the governmental institutions. Even in the arena of investigative commissions debates about videotex played a marginal role.

When business-sponsored Telebild entered the Swedish videotex scene, it moved away from the Prestel idea and instead developed a concept of a business-oriented videotex market. Since the commercial phase became delayed, this concept acquired further support in 1981/82. At that time Prestel's failure to create an expected videotex market based on households became a well-known fact. Therefore, Sweden could act as a "late-comer", benefiting from the 'mistakes' of the British (Gerschenkron 1986). This had as a consequence that the videotex developers in Sweden also rejected the Prestel-based concept of a terminal configuration. The redefinitions of videotex seem to constitute the chief reason for the relative success on the business market. But, they also explain its current stagnation and its "elitist" character contrasting with the more "democratic" French model.

The technical history of the Swedish videotex went through several turning points. Each meant a more or less radical break with the past. Each also had at its root the dissatisfaction of videotex IPs and IUs with the ineffectiveness of videotex.

As the Swedish videotex concept built on Prestel it had at its conceptual basis a hierarchical system architecture. The first turning point came in 1984 when Televerket decided to decentralize the system in order to make it more efficient. It added 2 computers and it worked towards introducing gateways between the main and external computers.

The second turning point, which brought this decentralization process to an end and gave the development of the Swedish videotex a new trajectory, came in 1985 when Televerket's top management suddenly decided to buy and adjust the IBM/BTX videotex system to the Swedish conditions. This decision caused a new wave of criticism directed at Televerket and contributed to the saturation of the videotex market. Apart from the access problem which decreased somewhat, the new technological import did not improve much the effectiveness of the videotex system. Although the IBM/BTX system had built-in vast interaction capacities, they remained underutilized.

The third turning point, which again breaks in a radical way with the previous developmental trajectory, came in 1987 when the first TeleGuide project was launched. TeleGuide stands for a partial departure from the business-oriented market concept and has an ambition to develop a technically advanced interactive communication system with a smart-card solution.

The central role of Televerket needs to be emphasized. It was the main decision-maker, net-owner and -provider, privileged IP, administrator, chief investor and researcher. Later on its roles became reduced to those of a net-provider, administrator and a more specialized decision-maker.

Governments role has been very limited. Unlike in France, Swedish videotex has never been conceptualized as a remedy against any industrial ill or economic stagnation. In contrast, videotex developers in Sweden did not count on or try to secure a share of the public budget devoted to economic modernization and expansion. Much more like in Germany or Great Britain, their resources were restricted to those provided by a public agency with self-financing status - Televerket, and an assorted range of business firms, the most capital intensive of which were gathered under the umbrella of Telebild.

Conclusion

Although the videotex developers in Sweden see the market as stagnating, their diagnosis differs markedly from ours. In their view, the greatest obstacle to the further expansion of videotex is its past orientation to a sophisticated user who has rejected videotex technology as both too simplistic and too costly.

In our view, in contrast, roots of the problem lie in the early choices of the videotex technology. The 1984-decentralization attempt and experiments with "quick and dirty" gateways became aborted when Televerket's top management decided to buy the IBM/BTX system in 1985. Contrary to its expectations, this purchase did not much improve the effectiveness of the system. Moreover, although the new system had vast interaction capacities, these remained underutilized. Not to be forgotten in this context is the fact that Televerket took unilateral decisions about technology, although on occasion AU-System and Telebild demonstrated a modicum of bargaining power. Along with its monopoly on pricing, in the long term these unilateral decisions undermined its own attempts at sustaining and expanding the Swedish videotex market.

The roots of the problem are also found in the fact that the developers of videotex have focused on the business market. Once this decision was taken, the service-provision has been accordingly tailored. This did not happen, however, soon enough to keep up the interest of the larger banks which were anyway unhappy with the inadequate solution of the security problem. The orientation to a business market, search for niches, and the informal sectoral agreements provided an initial impetus to the development of the "relatively successful" videotex market. It also secured a better statistics of subscribers than that of the actual users. But the provision of business-focused services caused disinterest in videotex among the wider public. Later on, when the search for further business-niches had come to a standstill, the very reason for the success of Swedish videotex turned into its "stagnation" sentence.

Whether a renewed attempt to revitalize and expand the Swedish videotex market, undertaken under the umbrella of TeleGuide project, will succeed, depends, in our view, not only on whether the attempt to reach the "unsophisticated" user will be successful. It also depends on the ability of the project promotors to institute the two-way, secure, and not too costly communication system, and, most of all, on their ability to make the videotex system efficient.

Future Perspectives

The question of the future of videotex has often been posed as in most of the countries the videotex appears not to be successful at all. But despite this failure of videotex in most of the countries, a new notion of a revival of videotex has emerged on the international scene.

The Swedish TeleGuide project, mentioned before, is parallelled by efforts in other reference countries to push the development of videotex. The project is seen as a way out of the saturation problems encountered in the Swedish business-oriented videotex market. The main attempt is not to find ways to expand it but to create a parallel, household-oriented market. Its penetration is considered a necessary prerequisite for a full break-through of videotex in Sweden.

The exclusive group of the early videotex enthusiasts, once more reshuffled as TeleGuide became incorporated, is now eagerly waiting for, what they call, the "ketchup effect". They believe that TeleGuide represents the last, decisive squeeze of the videotex ketchup bottle which will produce a big videotex-subscriber blob. This concept is much less sophisticated but akin to that of critical mass. Videotex developers' great hope is that the public-oriented and business-oriented services will mutually stimulate and reinforce each other and bring a wished-for real break through for the videotex market.

With a probable Swedish entrance in the European Market, and Europe's preparations for liberalization of telecommunications, it remains to be seen how a single European market in 1992 will change the diffusion of videotex.

References

Anton Tom. 1969. "Policy-Making and Political Culture in Sweden", **Scandinavian Political Studies** Vol. 4, pp. 88-102

Boman Mogens. 1983. **Teledata - framtidens infromationssystem**. Malmö. LiberFörlag

Domhoff G.William. 1979. **The Powers that Be**. New York. Vintage Books

Flam Helena and Rose Joanna. 1990. "**Introduction of Videotex in Sweden**". Research Report. Department of Sociology. Lund University.

Gerschenkron Alexander. 1986. **Economic Backwardness in Historical Perspective**. Cambridge. Belknap

Hughes Thomas P. 1987. "The Evolution of Large Technological Systems". In **The Social Construction of Technological Systems**. Edited by Wiebe Bijker, Thomas P.Hughes and Trevor Pinch. Cambridge: MIT Press, pp.51-82

Joerges Bernward. 1988. "Large Technical Systems: Concepts and Issues". In **The Development of Large Technical Systems**. Edited by Renate Mayntz and Thomas P.Hughes. Frankfurt am Main.Campus verlag, pp.9-36

Junberger Lars. 1983. **Teledata för Sverige**. Malmö. LiberFörlag

Katzenstein Peter. 1985. **Small States in World Markets: Industrial Policy in Europe**. Ithaca: Cornell University Press

Mayntz Renate and Schneider Volker.1987. "**Interactive Telecommunications: The Case of Videotex in Germany, France and Great Britain**". Paper prepared for the Conference on the Development of Large Technical Systems - Theoretical Approaches, Empirical Cases and International Comparisons. Cologne, Germany, November 25-28

Mills C.Wright. 1961. **The Power Elite**. New York

OECD. 1988. "**New Telecomunication Services. Videotex Development Strategies**". Report No 16, Paris

Ohlin Tomas. 1986. **Videotex**. Stockholm. Riksdataförbundet

Premfors Rune. 1983. "Governmental Investigations in Sweden", **American Behavioral Scientist** Vol. 26 No 5,pp.623-642

Riksbankens Jubileumsfond Reports:
RJ 1977:3. "**Hallå, hallå! Rapport från ett symposium om telekommunikationernas samhälls-roll**"

RJ 1978:5. "**Tryck på knappen! Om Viewdata och andra sätt för hushåll och kontor att tvåvägskommunicera med databaser**"

RJ 1979:1. "**I kulisserna. Om Viewdata/Datavision och andra medier som väntar på entré**"

Schneider Volker and Graham Thomas. 1988. "**The Policy and Politics of New Media: Comparative Analysis of Interactive Videotex in Great Britain, France and the Federal Republic of Germany**". Paper prepared for the mini-plenary session on International Communication and International Politics at the IPSA World Congress in Washington, D.C, 28.8-1.9

Schneider Volker, Jean-Marie Charon, Ian Miles, Graham Thomas, Thierry Vedel. 1990. "**The Dynamics of Videotex Development in Britain, France and Germany: A Cross-National Comparison**", paper prepared for the 8th International Conference of the International Telecommunications Society on "Telecommunications and the Challenge of Innovation and Global Competition" in Venice, Italy, March 18-21

Slaa Paul and Harry Bouwman. 1990. "**Videotex as an Intermediate Service**", paper presented at the Workshop on Videotex in a Comparative Perspective, Stockholm, May 18-19, 1990

SOU. 1979. Nya Vyer: **Datorer och nya massmedier - hot eller löfte**. Nr. 69. Stockholm. Utbildningsdepartementet

SOU. 1981. **Nya Medier: text TV, teledata**. Stockholm. Utbildningsdepartementet

Swedish Telecom. 1990. "**Facts about Televerket**", published by the Information Unit at Televerket Headquarters

Wittrock Björn. 1986. "**Development and Present State of Public Policy Research: Country Studies in Comparative Perspective**". Paper presented at the Swedish-German Social Science Project Workshop at WZB, Berlin, December

CHAPTER 11

IRELAND: FROM PUB TO PUBLIC

Gerard O'Neill
Henley Centre Ireland
Dublin

Introduction

Ireland is a relative newcomer to the videotex business, having seen the establishment of Minitel Communications Limited in 1990. The company is a joint venture among Telecom Eireann, Ireland's national telecommunications company, which has a 30% shareholding, AIB plc - Ireland's largest bank (20%), Credit Lyonnais (20%) and Intelmatique, a subsidiary of France Telecom (30%). Minitel in Ireland has been established with a similar brief to Minitel in France, namely to develop and promote videotex services for business and domestic customers through Telecom's National Videotex Access System (using an X.25 Packet Switched Data Network), and to distribute low-cost minitel sets which are fitted to user's telephone lines. The structure of the minitel system in Ireland is given in figure one.

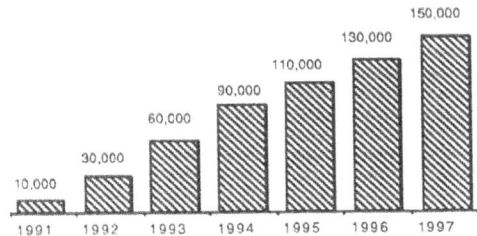

Figure 1: Projected Growth in Number of Minitel Sets placed in Ireland

(Source Minitel Ireland Projections)

The Minitel service in Ireland is only just emerging from the development phase, having gone 'live' on a fully commercial basis in March 1991, with the result that it is extremely early to be making prognoses about its eventual performance and future demand. Given this, we have focussed in this report on the issues, market-related and policy-related, which will shape the outlook for Ireland's first national videotex service.

We present below the results of research on the demand for videotex services, using comparative research between Ireland and the UK.

Current Situation

Ireland has experienced remarkable improvements in its telecommunications infrastructure over the past ten years. This has enabled Ireland to put in place sophisticated facilities for the provision of new networking, electronic mail, and related services. Thus, on the supply side, the means exist to provide the type of videotex services now planned. For a number of years, Ireland has had a Prestel type service available through RTE, the national television and radio

141

H. Bouwman and M. Christoffersen (eds.), Relaunching Videotex, 141–147.

organization. While this 'Aertel' service is now available in over 50,000 homes (1 in 20), it is not seen as a directly competitive system to Minitel, the latter having a major emphasis on transaction processing and not just 'information broking'.

A key, and very sensible, component of the Minitel strategy for Ireland is to encourage businesses to take up the initial services on the Minitel system, gradually extending the subscriber base to households. Businesses are less likely to be uncomfortable with keyboards and database type services - while households will have a slower learning curve.

By September 1991, following the formal launch of the service in April 1991, there were 700 subscribers with access to some 35 services - most of which are of a business-to-business nature. Of this total, some 200 minitel sets are actually located in Irish pubs. Publicans can use a service called Vintel, which allows them to process tedious and time consuming stocktaking requirements. Vintel has set itself a target of installing a minitel in 2,500 pubs out of Ireland's total of 12,500 over the next few years.

Minitel has set itself equally ambitious targets as shown in figure two, drawing on the experience of the French service in its early stages. It is unlikely, however, that they will achieve their first year target of 10,000 installed sets, given the quite depressed economic climate in Ireland during 1991, which has affected those considering supplying services as much as those considering acquiring a minitel set.

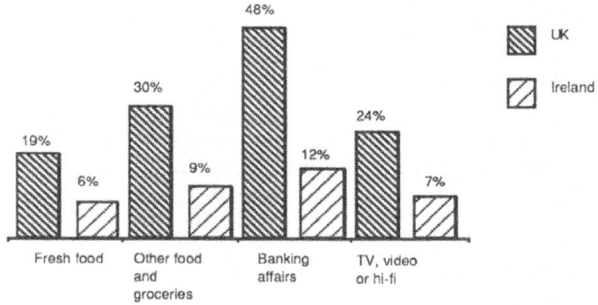

Figure 2: Home shopping. Proportion of adults interested in various home-shopping services: UK and Republic of Ireland
(Source: Henley Centre 'planning for Social Change studies', 1990-91)

Minitel in Ireland will have effectively spent the first two years of operation encouraging the growth in both the number of service providers and of bureaux specializing in the development of software to operate videotex services on behalf of service providers. Minitel in Ireland has decided to provide for all three videotex standards (Prestel, Teletel and ASCII). As a result it has been able to incorporate part or all of several pre-existing on-line services such as Dun & Bradstreet - helping quickly to build a 'critical mass' of services which will in turn attract a growing number of users. New services specially designed for the Minitel marketplace are gradually coming on stream.

Service providers are generally choosing between 5 tariff bands, with varying payback to providers - as set out in table 1.

Table 1. Tariff Bands in Irish Videotex System.

Tariff Bands applying in 1991	User Price Per Minute	Revenue to Service Provider*
1	0.0p	-6.5p
2	6.5p	0.0p
3	11.5p	5.0p
4	16.5p	10.0p
5	21.5p	15.0p

* the remaining revenue is divided 4.0p to Telecom Eireann and 2.5p to Minitel Ireland

By the end of 1991 it is expected that businesses and individuals with PCs and modems will be able to use specially developed software packages to access Minitel services, rather than acquiring a Minitel set as a present. That said, users will still be able to buy (or rent) Alcatel or Philips Minitel terminals, at Ir£250 (plus vat) for a standard set, or a rental charge of Ir£5 per month. In addition, there is a once-off registration fee of Ir£20 or Ir£50 for organizations with more than one screen.

Current users of the Minitel service in Ireland include a number of companies with private access to a discrete part of the network. One of these is Shell, whose commission agency petrol stations in Ireland are now keying in daily sales totals of each grade of petrol and autodiesel. Minitel sees these 'vertical networks' as important ways of introducing managers to the wider range of professional and personal services available on Minitel.

Policy Analysis

There has been little if any public debate about the policy implications of Ireland's first large scale development in videotex services. To some extent this is because there is a wide range of very advanced telecommunications services under development in the country, including EDI services. In addition, virtually all such services involve one or other of the key state companies in this field, ie: Telecom Eireann or An Post (the national postal service). Thus the 'public interest' is perhaps seen as being represented by these state-owned companies.

In Ireland, the attitude towards most telecommunications-based initiatives is that they are a 'good thing' for the simple reason that they will enable Ireland to overcome the problems of 'peripherality', which is seen as a major strategic issue in the context of the emerging single European market. Thus issues such as the implications for Irish society of information technology are rarely discussed by public representatives, and only occasionally touched on in debates about employment and unemployment. This clearly contrasts with the experience of other countries, such as Austria, which witnessed quite intense debates both before and during the promotion of a national videotex service.

Undoubtedly, the promoters of Minitel in Ireland have learned from the mistakes of other countries. Countries such as Austria found themselves in protracted and antagonistic debate (when developing videotex) among employers organizations, trade unions and government departments about developments that never lived up to their earliest promises (or threats, depending on which side of the debate you took). Videotex has not been 'sold' to the Irish people as a 'high-tech' solution to any of the country's ills, or as a key platform in a 'telecommunications-orientated' industrial policy. Instead it has had an extremely low key period of development - quietly encouraging service providers and would-be users, without the ever present threat of 'over promising' at too early a stage.

The Irish partners in the venture, Telecom Eireann and AIB Group, have deep pockets. They can afford not to get an early return on their investment - nor do they expect one. For Telecom Eireann, videotex is one of a range of new, related services under development, which are designed to complement the core telephony business. For AIB, it is a tool to use as part of an ever widening range of customer services, which portray the bank as being in the forefront of banking developments in Ireland.

Telecom Eireann states in its corporate plan for 1992/93-1996/97 that a primary strategic issue is 'how to position the Company to cope with an increasingly competitive environment for telecommunications'. That competition is seen as arising primarily from the relaxing of regulatory barriers to competition in national and international markets.

As a state-owned company, Telecom Eireann recognizes a strategic requirement to contribute to the national economy of Ireland. Thus the Minitel joint venture is seen as fitting into Telecom's desire to engage in any activity that can utilize the resources of the core business to generate profitable new business.

What remains unclear is how Telecom Eireann's interest in videotex services will sit alongside its current and growing involvement in cable television services (it owns 60% of Cablelink, Ireland's largest cable TV service), and in ISDN networks - likely to emerge over the next few years. Already it is clear that fiber optic cable systems and digital telephone lines will enable suppliers and users to access data transmission services of a quality that is far superior to videotex. Of course, videotex in Ireland has a major advantage, namely that it can be carried by existing analogue lines. Thus the threat from Telecom's other new business interests is still some way off, though the threat would first become apparent in the business-to-business sector which is the current focus of Minitel's strategy in Ireland. It should also be noted that the cable TV system uses co-axial cable with only limited interactive capacity - so considerable investment would be needed to upgrade the system for further service development.

For the French partners, Credit Lyonnais and France Telecom, Ireland has one key advantage that outweighs its small size, namely that it is English speaking. It is the first English speaking country to adopt the Minitel system on such a scale. Obviously, if it goes well, then other English speaking countries might also be targeted for bigger, and more lucrative developments. The fact, for example, that the UK's Prestel system has performed so dismally could be seen as an opportunity to start afresh with a Minitel system that has been tried and developed successfully elsewhere. And while this has not been stated either explicitly or implicitly as Intelmatique's strategy, there can be no doubt that the key to successful growth in any telecommunications-based business will lie in the development of international linkages and services that exploit the deregulated environment for the communications sector. Working in Ireland has provided valuable experience in dealing with a variety of partners, both PTT and non-PTT, as well as obviously dealing with an English speaking marketplace.

A forthcoming development will be the provision of selected French services to Irish subscribers using Minitelnet - a linking service to the French Minitel system. Though they have been selected because of their potential appeal to an Irish audience (eg: travel information), the fact that they are virtually all in French will limit the likely uptake of such services. That and the fact that the cheapest price French service will come in the highest price band for Irish services (ie: 21.5 pence per minute).

Given the very early stage of videotex development in Ireland, we do not have a 'story' to tell about the evolution of videotex, related public policy and customer demands for the service. We therefore consider it to be more appropriate to assess the developments that we consider to be key determinants of the prospects for videotex in Ireland - and indeed for other countries where videotex is a relatively new phenomenon. Here we draw on extensive Henley Centre research into social and technological changes that are shaping the forces for growth in videotex and other telecommunications services.

Our research has highlighted some of the problems that those intent on introducing highly technological services such as tele-shopping will face. The next chart compares the level of interest in selected home-shopping services in the UK and the Republic of Ireland (ie: the proportion of adults saying that they could imagine themselves tele-shopping for any or all of the listed services).

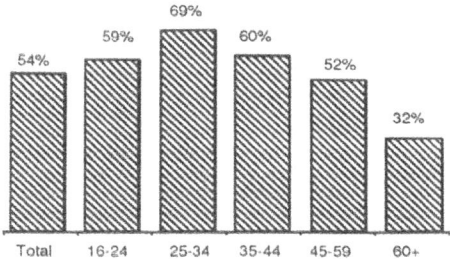

Figure 3: Home working: % of people in selected age groups who answer yes to the question: "If you had a job that paid you work from home would you do so?"

Across all categories there is a significantly lower level of interest in the range of potential services on offer in Ireland than in Britain. Moreover, even among the youngest age groups in Ireland, there is only a minority level of interest - though a larger minority than in the case of older age groups. Men are just slightly more interested than women.

Conclusion

It is too early yet to pronounce on the fate of videotex in Ireland. Minitel Communications Limited has had the opportunity to learn from the mistakes of other countries which have made expensive, and ultimately unsuccessful investments in national videotex services. These mistakes, highlighted in other country studies, can be summarized as:

- over-promising and then under-delivering - building up unrealistic expectations about what the service can do

- being over-focussed on 'state of the art' technology whilst ignoring the simpler needs of the marketplace

- letting technicians and engineers do most of the 'running' - with little or no regard for marketing and sales activities

- waiting until the subscriber base is 'large' before encouraging people to provide services

But of course videotex in Ireland faces a number of other hurdles that may not prevail to the same extent in other countries, including:

- a still generally low consumer interest in videotex type services

- a still low level of phone ownership among households

- the fact that the main bulk of international Minitel services are in French

- competition from other telecommunications networks also at a take-off stage in Ireland (eg: EDI and E-mail)

It is our view that an increase in home-working, related to wider changes in the marketplace, will play a crucial role in shaping the domestic demand for videotex and other interactive tele-communications services in Ireland. One technological development that is very much in the pipeline - if not already being practiced - is teleworking. In a study conducted for British Telecom in 1987, the Henley Centre estimated that, by the nature of their jobs, 14 million people could potentially be teleworking in the UK by the end of the 1990s - nearly half of all employees. Few of these people will do so full-time, but even if some only work for one day a week, new pressures will be placed upon the home in both practical and psychological terms.

There is a considerable level of interest in home-working in Ireland. The chart below shows responses to a question in our Planning for Social Change study which asked whether Irish respondents would be interested in working from home if paid to do so. Those age groups with young children in the home, 25-34 year olds, are most interested in home working. As they also fit the key profile of 'videotex pioneers' - ie: male, middle income, with children - they could be a key force in videotex development in Ireland.

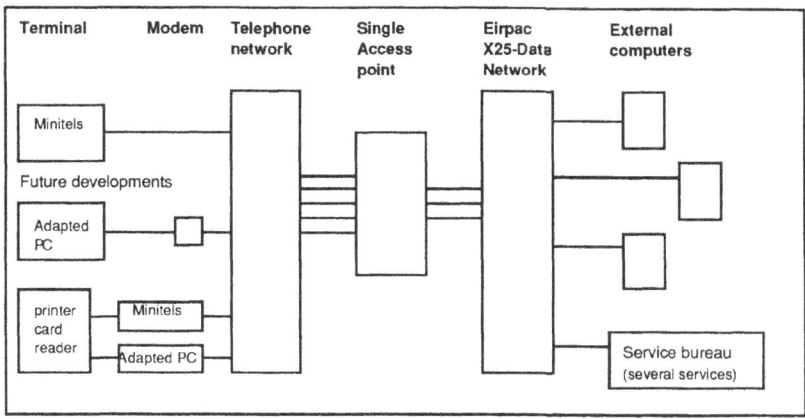

Figure 4: The Irish Minitel system

The prospects for videotex in Ireland will to a great extent be determined by both the evolving nature of telecommuncations services in general (eg: will ISDN make videotex redundant within ten years?), and by social change within Ireland - for example, will we see a situation such as currently in the USA where one in twentyfive households has computers linked to modems?

It is too early to answer these questions, but perhaps some of the answers will lie in what happens outside Ireland, just as the answers to many of the questions about our economic and political futures will also lie outside Ireland.

CHAPTER 12

US: VIDEOTEX IN A "HYPEREVOLUTIONARY" MARKET

Charles Steinfield
Department of Telecommunication
Michigan State University, Michigan, USA

The development of videotex in the United States both parallels and diverges from the experience of videotex in Europe. On the one hand, just as with their European counterparts, consumer acceptance of videotex services in the U.S. has not kept pace with overambitious projections. In addition, just as many European countries have continued to pursue videotex in the face of overwhelming consumer disinterest, US videotex ventures continue to appear despite several spectacular failures in the mid 1980s. On the other hand, unlike most European systems, there is no monolithic videotex network operator or service provider, but rather many competing ones, including several that are highly profitable and expanding rapidly. Also, there is not the same reliance on specific display protocols using special purpose terminals, as is more common in Europe. Instead, US videotex systems tend to rely on personal computers as terminal devices with ASCII text in place of graphics-oriented displays. In this chapter, a brief history of the development of videotex in the US is provided, along with a report on its current status. The distinct evolutionary path of videotex-like services is then shown to be in part influenced by the structure of the US telecommunications industry and corresponding policy and regulation with regard to telecommunications and information services. Several emerging policy issues raised by the continued development of electronic information services are also briefly discussed. The final section of the chapter introduces several new initiatives, in order to illustrate the almost "hyperevolutionary" characteristic of the electronic information services marketplace in the US. By hyperevolutionary, we imply that a sort of rapid natural selection process now appears to be unfolding, with many new forms appearing as experimentation to find successful services and modes of service delivery continues.[78]

Historical Overview

It is possible to trace back the genealogy of videotex in the US through several different branches defined by the key actors attempting to enter the interactive services market (see Case, 1990 for a social constructionist view of the history of videotex). Major forays into the videotex arena in the U.S. have at various times been pursued by the following types of actors:

- Publishing Firms (newspapers and specialized business-oriented publishers)
- Broadcasters
- Cable Television Operators
- Computer Manufacturers
- Computer Timesharing Services
- Data Base Services
- Telecommunications Firms

---

[78] Robert Pepper discribed the interactive services marketplace in a similar way in an interview published in the Information and Interactive Services Report, June 21, 1991.

H. Bouwman and M. Christoffersen (eds.), Relaunching Videotex, 149–164.
© 1992 *Kluwer Academic Publishers.*

PUBLISHING FIRM APPROACHES

Perhaps the first to begin exploring the feasibility of electronically delivered information services to businesses and homes were a few forward thinking publishing firms. Baer and Greenberger (1987) in their review of U.S. videotex efforts, report that as early as 1974, Dow Jones and Knight Ridder began exploring the market for electronic text services. Particularly for newspaper publishers, investigations directed at the home consumer were motivated by an attempt to protect their markets should electronic delivery of text prove popular. However, Dow Jones focused on the business market, and began offering in 1974 electronic access to business information services, including a range of financial services as well as the text of the Wall Street Journal. The financial success of the Dow Jones News Retrieval Service helped to convince many that ASCII-based text services, marketed toward the business user, represented the most viable approach to the electronic information service market. Other approaches by the publishing community were directed specifically towards the home market and were more influenced by the European view of videotex as a mass market service. Knight Ridder began test marketing their Viewtron system in Coral Gables, Florida in 1980, and in 1983 began offering it commercially. This system offered subscribers access to a range of information and transaction services, including home banking and shopping, news, sports, weather, stock market information, airline schedules, movie reviews, and so forth. They were convinced that home consumers needed graphics, and followed the Prestel philosophy of using the television set as a display terminal. A special decoder known as the Sceptre was developed for the trial by AT&T, using a graphics display standard based upon the Canadian Telidon system. The North American Presentation Level Syntax. (or NAPLPS) protocol, AT&T's revision of Telidon, provided higher quality graphics using an alphageometric approach, but required greater processing capability (and hence higher cost) in the decoders attached to television sets. It also was more difficult and expensive to create the pages of information to be accessed. The Sceptre was originally introduced into the marketplace at $900 (735 Ecus[79]), but the price was quickly reduced to $600 (490 Ecus). Even the reduced price was far too expensive for a dedicated terminal device for an unproven service. After three years, the service was shut down following nearly $50 million (40.8 million Ecus) in losses. Similar results awaited the Times Mirror Co, which lost nearly $30 million (24.5 million Ecus) in their Southern California based service known as Gateway. Both shut down within weeks of each other in 1986. An earlier venture started by Time, Inc., which explored the viability of offering a videotex-like service over cable using a full channel for delivery of pages, also shut down in 1983 after costing an estimated $25 million (20.4 million Ecus). These spectacular failures, and numerous smaller ones, had the effect of souring the US publishing community on the concept of home videotex. Perhaps more significantly, however, the failures convinced the publishing industry that videotex posed no real threat to their survival, at least in the near future. In retrospect, then, it seems clear to most observers that these trials were more motivated by a desire to protect an existing business than to successfully launch a new service (e.g. see Mansell, 1988).

BROADCASTERS

Most of the early attempts by broadcasters at providing electronic information services focussed on teletext rather than videotex. Both NBC and CBS developed national teletext magazines, which for the most part were either stripped off by local broadcasters, or were transmitted even though there were virtually no decoders in households. About the only "teletext-like" system that

---

[79] Whereever a figure in U.S. dollars is used, an approximation in European Currency Units at a rate of 1.225 Ecus to the dollar is provided.

was actually received by home users was ABC's Line 21 system which transmitted closed captions for the hearing impaired. CBS also dabbled in a few early videotex trials, teaming up with AT&T in one New Jersey trial, and participating as a partner early on in the formation of Trintex (which became Prodigy). The networks thus sought to ensure that the broadcasting community would play some role in the provision of electronic information services, but as with the publishers, only to protect their traditional service base, while possibly adding a new outlet for advertising.

CABLE COMPANIES

One of the earliest home interactive service ventures was the Qube cable system in Columbus, Ohio. The cable franchise owner, Warner, rolled out Qube with a great deal of fanfare, and incorporated versions of its interactive services in other franchise bids in its attempt to win new franchises in major markets (see Becker, 1987, for a detailed analysis of the Qube system). Interactive services included pay-per-view for movies and opinion polling. However, once HBO revolutionized the cable network business by starting national distribution of programming by satellite, most of the Qube channels that were dedicated to interactive services were changed over to standard television programming. Moreover, even though Warner was successful in winning several additional franchises, other Qube systems never materialized. The cable industry quickly lost interest in providing interactive services, claiming that they simply were not profitable. In retrospect, it appears clear that the interactive service promises were made simply to enhance franchise bids, and their early use in Qube merely helped to fill unused channel capacity until cable networks provided programming alternatives.

COMPUTER MANUFACTURERS

Many electronic information services were promulgated by computer manufacturers, in an attempt to encourage growth in home computer sales. An example of this type of service is QuantumLink, which provided a network for owners of Commodore computers permitting electronic mail, interactive games, and other types of transactions and information services. Several of these types of services have remained marginally successful, although by necessity, most have opened their services to IBM compatible and Apple computers as well. Computer vendors have continued to play a role in new videotex ventures (e.g. IBM and Prodigy, discussed in a later section.).

COMPUTER TIMESHARING SERVICES AND VALUE ADDED NETWORK PROVIDERS

Among the most successful entrants into the electronic information services marketplace were the computer timesharing firms. Timesharing firms, which offered remote computing services to businesses expanded to the electronic information services business as demand for remote computing diminished with the proliferation of computers in business. Importantly, all of these types of services were ASCII text scrolled to computer terminals or personal computers. Among the most successful of these firms is CompuServe, a Columbus, Ohio based firm that was one of the earliest to enter the home market for information services in 1979. As they remain a profitable player in this area, more detail on CompuServe is provided in the section on the current situation.

An example of a value added network provider entering the home information services market is GEISCO (General Electric Information Services Company) which sought to take advantage of unused capacity on their network in the evenings. They created GEnie, a network for home computer owners that now competes with CompuServe and Prodigy in the videotex market.

Traditionally, electronic data base providers made their products available only through intermediaries, who facilitated searches in order to help minimize the rather expensive costs of access. However, several data base companies have offered direct access, primarily to professional users over one of the many packet switched networks available in the U.S. Prominant, and generally profitable services include the Lexus/Nexus/Medus databases offered by Mead Data Central, and the voluminous Dialog service, which offers access to hundreds of specialized databases. Once again, ASCII text is delivered to computer terminals or personal computers in all of these types of services.

TELECOMMUNICATIONS FIRMS

There are many different types of telecommunications firms in the U.S., many of which existed even before the breakup of AT&T. They all have had distinct roles to play in the development of the information services market, with specific initiatives greatly constrained by regulatory policy (more will be said on this in a later section). AT&T has through the years attempted to foster growth in this area, but up until very recently was completely prohibited from providing content, and could only serve as the network provider. In addition, they could also play a role in manufacturing of terminal equipment through their Western Electric subsidiary. Thus, AT&T always teamed up with an information provider in their early videotex efforts (e.g. KnightRidder with Viewtron, CBS in the Ridgewood, New Jersey trial), and sought to promote their NAPLPS display standard and Sceptre terminal. Their chief competitor in the long distance market, Microwave Communications, Inc. (MCI), offered a national public electronic mail service, primarily to professional users, but sought to enhance this service by comarketing it with access to the Dow Jones News/Retrieval Service.

The newly created Regional Bell Operating Companies (RBOCs), were also completely restricted from providing information services for fear that they would abuse their position as the sole provider of the local access network. Regulators and competing information service providers argued that not only could the RBOCs use revenues from their regulated services to subsidize information services, but also they could impose other disadvantages on other information service providers (such as higher costs for billing services, lower quality connections, etc.). In 1988, they were allowed to enter this market in a more limited way, with permission to offer specialized gateway services for providing access to independent information service providers. A more recent decision has seemingly eliminated information service provision restrictions, but remains under appeal at the time of this writing.

In general, unlike the PTT dominated videotex services in Europe, telecommunications firms, particularly those affiliated with the former Bell System, have been prohibited from offering information services. Thus, their role has generally been as a partner providing the transmission services, or in the case of AT&T, terminal equipment. A cynical view on telephone company participation in videotex (particularly the RBOCs) is that they have no real interest in seeing the market flourish under the present regime of line of business restrictions. Continued difficulties in developing the market can then be blamed on these restrictions, serving to strengthen telephone company arguments for relief.

Current Videotex Situation

Given the many different paths into the electronic information services market, it is not surprising then that the U.S. videotex situation is very complex. Because of the many different influences, there is no single definition of what constitutes videotex. It includes all easy to use electronic information services that do not require specialized computer skills for use. Scrolling ASCII text services such as CompuServe are considered as much videotex as graphics-oriented services such as Prodigy. The borders, then, between videotex and more traditional online services are quite blurry. We thus present the current videotex situation within the context of a larger online services industry that exists in the U.S.

One factor that differentiates the U.S. from Europe is the sheer size of the online market, which has been estimated by one consulting firm to have nearly $9 billion (7.35 billion Ecus) in revenues for 1990 (Table 1). Note, however, that financially-oriented services to the professional community dominate, accounting for nearly 85% of the market. Home consumer services constituted only 2.7% of the online services market according to this estimate, although they were described as being the fastest growing segment. Moreover, although small in comparison to the total online market, revenues from just one consumer service provider, CompuServe, was over $200 million (163 million Ecus), reportedly approaching the revenues of the entire Minitel information provider community (estimated at about $261 million [213 million Ecus] in 1989, Kramer, 1991).

Table 1. The US Online Services Market 1990

- 4,200,000 online service subscribers
- 90% growth rate in revenues from 1986 to 1990
- 1990 revenues of nearly $9 billion
  (7.35 billion Ecus)

Market Segments

Financial/Brokerage	49.0%
Credit Reporting/Verification	18.2%
Financial News/Research	16.1%
Legal/Regulatory	7.5%
Professional	6.5%
Home Consumer	2.7%
Marketing	0.1%

source: SIMBA Information, Inc. cited in Information and Interactive Services Report, June 7, 1991.

The consumer videotex market is quite small relative to the size of the potential using population which could include approximately 90 million households (see Table 2 for estimates of the number of subscribers to major consumer services). The major services have, however, experienced fairly steady growth, as evidenced in Table 3, which shows the number of subscribers for a few of the services in the years 1987, 1988, and then again for 1990.

Table 2. Major Consumer Videotex Services in the United States

Service Name	System Operator	Number of Users
CompuServe	CompuServe (H&R Block)	760,000
Prodigy	Prodigy Services Co. (IBM and Sears)	550,000
GEnie	GE Information Services Co.	252,000
PC Magnet	Ziff Communications Co.	120,000
CUC Online	CUC International	81,000
Delphi	General Videotex Corp.	80,000
PC-Link	Quantum Computer Serv. (with Tandy)	60,000
QuantumLink	Quantum Computer Serv. (with Commodore)	60,000
America Online	Quantum Computer Serv.	54,000

Source: Arlen Communications, March, 1991

Table 3. Growth in Subscribers for Selected Services

Service Name	Number of Users		
	87	88	91
CompuServe *	340,000	416,000	760,000
GEnie	37,000	85,000	252,000
Delphi	45,000	55,000	80,000

* includes merger with the Source (70,000 subscribers in 1988).

Source: NTIA Telecom 2000 Report (Matos, 1988) for the 1987 and 1988 tables, Arlen Communications for the 1991 tables. 1991 tables are identified as being as of March 1. The specific month for the 1987 and 1988 tables are not identified in the NTIA report.

COMPUSERVE VS. PRODIGY

As seen in Table 2, the two largest consumer videotex operators in the United States are CompuServe and Prodigy. A brief profile of each will illustrate key characteristics of US videotex approaches.

CompuServe, formerly a computer timesharing firm, opened the CompuServe Information Service in 1979. In order to use CompuServe, a subscription is required, as well as a microcomputer or computer terminal, a modem, communications software, and of course, a telephone. A subscription costs $39.95 (32.6 Ecus) for a kit that includes the user ID, a communications software package designed to work with CompuServe, and $25 (20.4 Ecus) worth of usage credit. A usage charge is a function of the connection time, and the price varies according to the transmission speed of the connection ($6.00 [4.9 Ecus] per hour for 300 bps, $12.50 [10.2 Ecus] per hour for 1200/2400 bps, and $22.50 [18.4 Ecus] per hour for 9600 bps). CompuServe offers no graphics - it is strictly ASCII text. As of September, 1991, CompuServe had 841,000 subscribers worldwide, including about 100,000 subscribers to their Japanese subsidiary, NiftyServe. It is striving to be an international operator, and, in fact, does not release subscriber information for domestic and international users separately. CompuServe operates its own extensive packet switched network, enabling a local call from over 97% of the U.S. It also maintains nodes in over 100 countries. There are approximately 1500 different databases and services available on CompuServe, with approximately 600 provided and maintained by CompuServe, and 900 accessible through gateways to other service providers. Most popular are the communications-oriented services, including bulletin board, conferencing, and electronic mail services. The highly used Special Interest Group Forums, which are asynchronous conferences focusing on specific topics, include user groups for various types of personal computers. Also highly popular are the downloading services for public domain software kept in an online library. Other types of services available include news, weather, and sports information, as well as entertainment, education, research (e.g. database access), travel, shopping, and financial services. CompuServe used to offer a separate business information service and a consumer information service, but now the two are merged, as they discovered that the same types of services were used by both types of subscribers. They do, however, market private (i.e. closed user group) videotex and electronic mail services to companies for intra and inter-organizational transactions and information services (for a case study of a firm using CompuServe in this way, see Steinfield and Caby, 1990). Finally, as shown in Table 4, CompuServe has established a solid record of profitability, contrary to the results of many other consumer videotex offerings. It now estimates that new subscribers are being added at the rate of 10,000 per month. To help compete with its new competitor, Prodigy, CompuServe acquired another leading information utility, The Source, in 1988.

Table 4. CompuServe Revenues and Pre-Tax Earnings, 1985-1991

Year	Revenues		Pre-Tax Earnings	
1985	$68,862,000	(56.2)*	$7,153,000	(5.84)
1986	84,855,000	(69.4)	10,000,000	(8.16)
1987	102,855,000	(84.0)	15,494,000	(12.65)
1988	130,706,000	(106.7)	20,933,000	(17.09)
1989	173,116,000	(141.3)	30,620,000	(25.00)
1990	206,730,000	(168.8)	40,349,000	(32.93)
1991 1st quarter only	65,800,000	(53.7)	12,100,000	(9.88)

Source: CompuServe Information Services

* Figures in parentheses represent millions of Ecus

The Prodigy Services Company began as Trintex, a partnership between IBM, Sears, and CBS. CBS soon withdrew, and the name was changed to Prodigy. It began operations in the summer of 1988, after reported investments of approximately $250 milllion (204 million Ecus) by Sears and IBM (Matos, 1988). Prodigy initially offered services in selected markets using its own leased network consisting of an MCI backbone to local IBM Series 1 computers. After June of 1990, nationwide access to Prodigy was available through Tymnet, a large packet network now owned by British Telecom, as well as through a number of other regional packet network providers. Prodigy officially commenced its national marketing efforts in September of 1990. Although they do not release subscriber information, they appear to be adding users rapidly, increasing from 170,000 in 1990 (reported in Case, 1990), to 550,000 as of March 1, 1991 (Arlen, 1990), to a million in June, 1991 (from a source at Prodigy). The distinction between "user" and "subscriber" is important, however. Each subscriber is allowed to enroll up to five additional users; generally other family members. Prodigy counts the number of enrolled users, rather than subscribers, because of the way in which they generate revenue. Rather than rely on a usage sensitive tariff, Prodigy charges a single monthly rate of $12.95 (10.97 Ecus) per subscription, entitling all users on the subscription to unlimited usage of virtually all Prodigy services. A subscription kit containing special software needed to access the service costs $49.95 (40.78 Ecus). Prodigy derives revenue from advertising, and maintains that the actual number of users, rather than subscribers is the more important table on which to base advertising rates (not unlike other advertiser-supported media). Advertising industry sources have expressed strong satisfaction with the highly targeted approach that Prodigy permits, as ads can be selectively used depending upon the demographic characteristics of each user that signs on (Detroit Free Press, August 12, 1991). In addition, purchases made by users result in a commission to Prodigy, so that the actual user number again is relevant.

Prodigy uses a combination of NAPLPS and ASCII. To hold down network costs, which ordinarily would be higher with graphics, Prodigy requires either an IBM compatible or Macintosh personal computer with a hard disk and modem. The special software provided with the startup kit then uses the hard disk to store graphic templates of the accessed screens, transmitting only the necessary ASCII updates. Over 850 services are offered, all by externally contracted providers. Prodigy Services Company then packages the external services into a consistent format. Services include a wide range of information, electronic mail, education, entertainment, travel, financial, and shopping applications, and there is a heavy focus on family oriented applications. The unlimited usage enabled by the flat fee thus encourages more time online by users, while enhancing the advertising value of the medium.

No financial information is available from Prodigy, except that officials estimate that a breakeven on investment will be reached in the early 1990s. Annual investments by IBM and Sears peaked last year and are now declining. Finally, a recent move has been to tap into the business services market, with the creation of Prodigy Business Services. The primary motivation was to take advantage of unused capacity during the day, exactly the opposite situation from operators who first targeted offerings to the business community.

A common aspect of videotex in the U.S., illustrated by the above profiles, is the near total reliance on personal computers as the terminal device. Although efforts have been made in the past to use special decoders with television sets, and more recently to import Minitel terminals, none of these have attracted more than a handful of users. Given the installed base of personal computers, which now is believed to exceed 20 million units and include 23% of all U.S. households (Kramer, 1991; see Table 5), any attempt to develop a service not based upon personal computers seems foolhardy. With the tremendous decline in modem prices (now available for under $100 [81.63 Ecus]), most new units are shipped with modems, and estimates of modem penetration among computer owners range from 25% to 50%.

Table 5.   **Personal Computers in U.S. Households**
                    Based On Units Shipped to Home Markets

Year	Cumulative Units	
1981	149,000	
1982	1,897,000	
1983	5,252,000	
1984	8,682,000	
1985	10,872,000	
1986	13,162,000	
1987	15,212,000	estimated
1988	17,222,000	estimated
1989	19,197,000	estimated
1990	21,107,000	estimated

Source: NTIA Telecom 2000 Report (Matos, 1988)

A second generalization that can be made about U.S videotex is the trend towards "flat rate" (i.e. unlimited use for a set monthly fee) pricing structures (Kramer, 1991) Table 6 lists the major consumer services that use such a pricing strategy, and even CompuServe has felt compelled to initiate a flat rate test. Their "basic services test" made its debut in the summer of 1991 among a test group, and provides unlimited use of a core group of services for $7.95 (6.49 Ecus) per month. Thus, these services no longer have the built in disincentive for use that appears to constrain Minitel users (Kramer, 1991), but requires that revenues be made up through other means such as advertising and transaction commissions.

Table 6. Consumer Services Using a Flat Rate Pricing Structure

Service	Cost per Month	
Prodigy	$12.95	(10.57)*
PC Link	9.95	(8.12)
QuantumLink	9.95	(8.12)
GEnie	4.95	(4.04)
CompuServe **	7.95	(6.49)

Source: Adapted from Arlen Communications, March, 1991
* Figures in parentheses represent Ecus
** recently introduced for a core set of services

RBOC GATEWAY SERVICES

Another group of recent entrants into the videotex marketplace in the U.S. are the Regional Bell Operating Companies. Until 1988, RBOCs were prohibited from any involvement in the provision of information services except, of course, for the basic local transmission lines. A modest relaxation of this prohibition came in 1988, when RBOCs were permitted to offer gateway services. Through a gateway, users could dial one number and be provided with an RBOC-maintained menu of available information services. They could then choose the desired service and be connected. Surprisingly, many RBOCs actually charged a subscription fee for this service, resulting in higher costs to the user than if they went directly into the information service. Given the relatively little added value of gateways, and the lack of participation by several major consumer videotex providers - including Prodigy and CompuServe, who maintained their own networks - the RBOC gateways experienced little success (see Table 7). In fact, Nynex recently announced that they were discontinuing their Info-look gateway services.

Table 7: RBOC Gateways

Service Name	RBOC	Subscribers
Intelligate Philadelphia	Bell Atlantic	7,000
Intelligate Washington	Bell Atlantic	7,000
Transtext Universal Gateway	BellSouth	3,000
Burlington Info-Look	NYNEX	500
Boston Info-Look	NYNEX	6,000
New York Info-Look	NYNEX	7,000
Community Link	US West	5,800

Source: Adapted from Arlen Communications, March, 1991

Although the US market appears quite fragmented, there is actually a complex web of interconnections across the various services. In effect, most videotex operators function like gateways themselves, providing access to numerous external information service providers in addition to their own internally provided services (see Jackson, 1990 for a detailed discussion of the gateway concept). For information service providers, a critical mass of users is reached by establishing links to multiple videotex system operators. Thus, the Official Airline Guide (OAG) can be accessed directly, or through CompuServe, Delphi, Dialog, Dow Jones, and so forth (Jackson, 1990). Two-step linkages into OAG are possible, as MCI mail subscribers, for example, can first link to Dow Jones, and then access OAG. Many services also provide gateways into Dow Jones, Dialog, and other popular information services and databases (See Table 8 for a listing of several popular services that function in this way). Thus, for information providers, the market is not necessarily so fragmented as it first appears. Recent linkages among formerly disparate electronic mail systems (e.g. MCI mail and CompuServe) further facilitate the building of a critical mass of users.

Table 8: Major Professional Information Services Potentially Accessible Through Consumer Videotex Systems

Service Name	System Operator	Number of Direct Subscribers
Dow Jones News/Retrieval	Dow Jones & Co.	350,000
Lexis/Nexis/Medis	Mead Data Central	270,000
Dialog	Dialog Information Services Inc. (Knight-Ridder)	120,000
OAG/EE	Official Airline Guides (Maxwell Communications)	40,000

Source: Arlen Communications, March, 1991

Policy Analysis

The structure of the U.S. videotex market is linked to the history of regulatory policy in the tele- communications sector in complex ways. Historically, information services have been provided on an unregulated basis by entities other than the common carriers subject to rate regulation under the Communications Act of 1934 (Matos, 1988). Regulated common carriers historically were not permitted to offer information services for fear that they would abuse their control over the network and engage in anti-competitive behavior against other information providers. Because the question of whether or not regulated common carriers can provide information services has become extremely sensitive, it forms the central focus of this section. However, because of space requirements, we attempt to highlight only the key developments, in order to illustrate how the videotex market has been affected.

Perhaps the early seeds of the current videotex situation were laid in 1956, when AT&T entered into a Consent Decree with Justice Department to settle an earlier anti-trust suit. In this decree, AT&T agreed to restrict itself to the provision of only regulated telecommunications services, thus foreclosing entry into aspects of the computer industry. These restrictions, coupled with a series of decisions that 1) permitted non-Bell equipment to interconnect with AT&T's network, and 2) resale of bulk capacity purchased from AT&T, allowed the growth of independent value added network providers such as Telenet and Tymnet (formerly known as Specialized Common

Carriers). Many of these SCCs formed the network infrastructure for future videotex network operators. Moreover, they aided the growth of the online services industry, because of arguably lower costs for network services than European counterparts which were forced onto a national X.25 network that faced no competitors.

More recent policy developments are focused specifically on the question of telephone company provision of information services. Initiatives originating from two distinct actors in the policy area are important here. First, for more than 20 years, the Federal Communications Commission has attempted to specify the conditions under which regulated common carriers could provide unregulated information services. These efforts are mostly centered in a series of actions known as the Computer Inquiries. The second major actor is U.S. District Judge Harold Greene, who presided over another antitrust suit first brought against AT&T in 1974 and resolved through a Consent Decree in 1982 that resulted in the breakup of AT&T (see Matos, 1988 for a more detailed discussion of information services policy in the U.S).

THE FCC COMPUTER INQUIRIES

In three major inquiries, the FCC has attempted to define the conditions under which regulated common carriers could provide unregulated, "enhanced" services. In Computer Inquiry I, the FCC decided to analyze on a case-by-case basis whether a service provided by a carrier under its jurisdiction was more like communications or more like data processing (Matos, 1988). Those that were considered to be data processing could only be offered under strict separation from regulated, tariffed services. This regime soon became unwieldy, however, and in Computer Inquiry II, the FCC attempted to define two classes of services, basic transport - in which the content was not modified in any way, and enhanced services - in which some processing of content did occur. The FCC ruled that AT&T, and subsequently the divested RBOCs, could only offer enhanced services through a structurally separate subsidiary. Distinguishing an enhanced service from a basic service has proven to be more formidable in the age of digital networks, and yet a third Computer Inquiry was held. In this final inquiry, the FCC decided to eliminate the structural safeguard approach, and substitute a new set of safeguards built upon accounting and access controls. Initially, the principle of Comparably Efficient Interconnection was proposed, which would require AT&T and the RBOCs to offer the same type of network access to external information service providers as they provide to their own information services. Additionally, the Open Network Architecture principle was proposed, which requires carriers to unbundle their network services and offer basic service elements to information providers on a tariffed basis (Matos, 1988).

The Comparably Efficient Interconnection and Open Network Architecture plans of Computer Inquiry were called into question in the summer of 1990 following a court ruling that the FCC had not adequately demonstrated that accounting controls could indeed ensure that the RBOCs would not engage in cross-subsidization from regulated to unregulated services. However, the FCC maintains that an Open Network Architecture approach is the best alternative to foster growth in information services, and has continued to request that RBOCs submit their plans for offering these types of services..

Other FCC actions have illustrated their desire to foster the growth of a competitive information services industry in the U.S. Significantly, they have continued to exempt enhanced service providers from having to pay access charges for their use of public network (Matos, 1988). Access charges were paid by interexchange companies to compensate local exchange providers for the use of their networks after the AT&T divestiture. In fact, an FCC proposal to require access charges of enhanced service providers was vehemently opposed by representatives from information industry, who were eventually successful in their lobbying efforts.

Virtually all of the FCC action in the area of information services has been preempted by the historic developments following the 1974 Justice Department anti-trust suit against AT&T. The culmination of this case was another Consent Decree filed by AT&T in 1982, eventually resulting in the Modification of Final Judgment (now universally referred to as the MFJ by industry observers) by District Court Judge Harold Greene. Because of the continuing oversight of the MFJ by Greene, he has significantly shaped the information services industry in the U.S. The MFJ resulted in AT&T's divestiture of 22 local Bell operating companies, which were reorganized into the present 7 RBOCs. AT&T maintained its long distance business, AT&T Bell Laboratories, and its manufacturing arm (formerly Western Electric). In addition, AT&T was freed to enter into other lines of business, with the specific condition that it could not enter the electronic information services business for 7 years. Three line of business restrictions were placed on the newly separated RBOCs. Specifically, they were prohibited from 1) providing long distance services, 2) engaging in any manufacturing activities (e.g. of customer premise equipment), and 3) offering any information services in which they would function as content provider.

Since the actual divestiture, which was carried out in 1984, the RBOCs have sought to weaken the restrictions placed upon their activities by a series of waiver requests and appeals. These requests are dealt with singlehandedly by Judge Greene, who in effect, has preempted the ability of the FCC to determine the conditions under which regulated common carriers might provide information services.

Over the years, a gradual loosening of the restrictions has occurred, but for the most part, Greene remained convinced that the RBOCs would use their monopoly control over the local access network to favor their own information services at the expense of others. In 1988, he permitted the RBOCs to offer gateway services, although they could not offer their own content. As noted above, the gateway services were all failures in the marketplace, and the RBOCs continued to pressure for full permission to enter the information services market. Then, after a federal court of appeals questioned his maintenance of the information services restriction on the RBOCs, Greene lifted the restriction in the summer of 1991. However, he attempted to delay the implementation of his decision (in effect, he maintained the prohibition) until all appeals had been exhausted. Naturally, major competitors such as CompuServe and Prodigy did appeal the decision, claiming that there is still not a "level playing field." At time of writing, however, Judge Greene's stay was determined by another court to be invalid, and the RBOCs have now been tried to enter the information services market. In fact, many observers feel that even if the lifting of the restriction is upheld, the RBOCs will still not invest the required resources to establish successful consumer information services, preferring instead to blame their continued market failures on the remaining line of business restrictions. RBOCs note, for example, that the long distance restriction would require the placing of a videotex access point in each local access and transport area (known as a LATA, which denotes the area in which a local exchange provider is allowed to provide services under the MFJ; interlata services are provided by interexchange companies such as AT&T and MCI), which may not be the most economical approach. They will probably withhold major investment in consumer videotex until all such restrictions are lifted.

The 7 year moratorium on AT&T's entry into the electronic information services was lifted in 1989, but the company has yet to respond with a major consumer videotex venture. Some observers feel that a few significant regulatory impediments remain, noting, for example that the expected offering of an electronic yellow pages by AT&T did not materialize because the specific authorization for a directory service did not include the ability to provide advertising (de Fontenay and Savin, 1990).

Thus, although the information services market in the US. is characterized by extensive competition among a variety of operators and information providers, regulatory policy has resulted in minimal telephone company involvement other than as a provider of transmission capacity with relatively little added value. The key question now, of course, is whether entry as a provider of information services by AT&T and the RBOCs would drive out existing information service providers, or result in a more robust information services.marketplace for everyone.

Another policy controversy arising from the continued debate over common carrier provision of enhanced services is the emergence of a conflict between federal and state regulators regarding who has jurisdiction. State public service commissions, which regulate intrastate telecommunications, argue that they have the authority to regulate carrier provision of enhanced services within their state, particularly if provided over a carrier's own facilities (Information and Interactive Services Report, August 30, 1991). Meanwhile, a bill in Congress is attempting to ensure that the FCC will have full jurisdiction over the enhanced services issue.

Finally, many other policy dilemmas are unfolding, even if they are somewhat overshadowed by the events related to the MFJ. Two issues with profound implications for information services provision are the privacy of interpersonal communications on privately owned systems, and the copyrightability of databases. In the former issue, several cases of a private videotex operator censoring private communications on their systems have raised the eyebrows of concerned observers (IISR, April 26, 1991). On Prodigy after a highly publicized dispute surrounding the imposition of a new charge for sending more than 30 messages per month, several people who conducted an online campaign against the service had their subscriptions cancelled. On GEnie, a person accussed of using profane language was censored and ultimately prohibited from using the system after continued confrontations with system operators. In another Prodigy controversy related to privacy, the operator was accused of intentionally examining users' private files on their hard disks. These charges were denied by representatives of Prodigy, who suggested that private data found in the disk space allocated for Prodigy use were simply deleted files that had not yet been overwritten. News of the charges circulated so quickly and widely that Prodigy felt compelled to submit to an outside security audit by Coopers and Lybrand, and to offer at no cost a utility that ensures that only Prodigy data can appear in the disk space set aside for the service's use (IISR, August 2, 1991).

The latter issue, which potentially impacts all providers of electronic databases, resulted from a Supreme Court decision early in 1991 that ruled that a telephone company did not have copyright protection on its telephone directory (US Supreme Court, 1991). They had sued a firm which had copied names for use in its own directory after the telephone company refused to sell the listings. The majority opinion of the court was that the compilation in quetion had "insufficient creativity" to justify copyright protection, and rejected the notion that the degree of effort necessary to compile the directory was enough. Thus, the question of whether electronic databases are protected by copyright now seems to hinge on owners' abilities to demonstrate that enough creative effort was involved to make the database truly an original production.

New Trends

The final chapter on U.S. videotex is by no means written. Despite the perception by many that videotex is a complete failure in the U.S., there still appears to be continued dynamism in this area, as evidenced by the almost weekly announcements reported in the trade press. Potential competitors in the videotex and information services marketplace include (but are not limited to):

AUDIOTEX AND DIAL IT SERVICES.
With the 900 and 976 dialing prefixes, thousands of information services are provided using voice storage and retrieval technology. Kramer (1991) estimates that the

combined revenues of audiotex and dial it services in the U.S. exceeded $2.5 billion (2.04 billion Ecus) in 1989. He notes, however, that like the early days of Minitel, the image of audiotex has been clouded by highly visible and controversial pornographic services.

COMPUTER BULLETIN BOARDS.

Reportedly numbering in the tens of thousands (Matos, 1988; Kramer, 1991), bulletin boards proliferated rapidly throughout the 1980s. Usually local in nature, and free of charge, bulletin boards offer users public and private communications and public domain software exchanges. Many professional and government services are now made available to users over bulletin boards. One noteworthy system, known as the Cleveland Free Net, began by providing access to medical expertise and subsequently added access to a range of government services. The model was considered so successful that other cities throughout the U.S. have begun installing the same system (Matos, 1988).

FAX NEWSPAPERS AND ELECTRONIC CLASSIFIED ADS.

A number of news publications, in their continued efforts to find a viable approach to electronic publishing, have attempted to capitalize on the facsimile boom by offering abbreviated, customized, or early versions of their papers by fax. In another interesting development, several newspapers have agreed to offer electronic classified ad services on Prodigy and other videotex systems (IISR, August 2, 1991). The protection of classified ad revenue was considered to be a primary motivation for many earlier and unsuccessful videotex ventures by the newspaper publishers.

VIDEO GAME NETWORKS.

Yet another potential pathway to home videotex services may come from the huge installed base of video games. Nintendo and AT&T have been rumored to be planning a video game network that would include other types of information services (Kramer, 1991). Another major video game vendor, Sierra Online, has set up The Sierra Network (TSN) using the Telenet network (IISR, May 24, 1991).

MINITEL SERVICES.

Minitel Services Company, a subsidiany of France Telecom, has been operating in New York for several years, and is accessible nationwide through Infonet, a packet-switched network operator. All french Minitel services are accessible. More recently, France Telecom and US West formed a $75 million joint venture to offer "Minitel like" services, beginning in Minneapolis, and using special purpose terminals.

SMARTPHONES.

In an attempt to recapture some role in the home videotex terminal market, AT&T plans to introduce the Smartphone, a combination telephone/video screen device, as part of a home banking service offered by an Ohio bank (IISR, August 2, 1991). Initially offered at $400 (326.5 Ecus), plans are to reduce the price to $200 (163.3 Ecus) within two years.

NATIONAL GATEWAY SERVICES.

Despite the relative unpopularity of RBOC gateway services, a new nationwide gateway service was launched by the National Videotex Network Corporation (NVN) (IISR, June 7, 1991). NVN is using AT&T's packet network to provide a common gateway to information providers. Many of the services can be accessed free with the core subscription of $4.95 (4.04 Ecus) per month, but others will require connect time surcharges. It is a multiprotocol gateway that offers access to services in NAPLPS,

ASCII, and Teletel system formats, and plans a connection to Minitel services through the Minitel Services Company.

RESIDENTIAL MULTI-TENANT SERVICES.

An interesting mix of telecommunications, cable, real estate, and financial executives have formed a new company called Intelesys, which will offer so called "residential multi-tenant services" (IISR,June 7, 1991). Essentially, they will act as a reseller of basic and enhanced telephone services primarily to apartment complexes using a shared PBX.

INTERACTIVE TELEVISION.

Not wishing to be counted out of the interactive services market, broadcasters are making a late push through a new radio-based interactive technology known as the TV Answer System (IISR, June 7, 1991). With TV Answer, a low power transmitter on the television set top sends digital information, including household ID and potentially alphanumeric data, back to a broadcast station headend. Originally touted as a pay-per-view technology that can also enable opinion polling, it is now being thought as a delivery mechanism for home shopping and other interactive information services.

Conclusion

In summary, the U.S. is likely to experience intense competition between many different forms of electronic information services. Although it appears that the ASCII text, personal computer-based systems will drive the home and business videotex market, other technological approaches, especially audiotex, may function more as mass market services. Also, unlike European systems which have many information providers accessing a common network, and integrated by one videotex system operator (usually the PTT), U.S. telecommunications policy has resulted in many competing videotex system operators, with some using their own networks, and others leasing network access from one or more of the many different network operators. In general, most system operators provide some core services directed at their particular market segment, but will add value to their service by providing gateways to other information providers. In this way, information providers will be accessed by many different videotex system operators, effectively broadening their market access and helping to achieve the necessary critical mass of users. Critical mass of information services and users can only be viewed in this context of segmented but interconnected systems. In many ways, the interconnected segmentation approach adds value to the users by enabling more targeted provision of core services without sacrificing access to potential mass market and highly specialized services as well.

References

Arlen, G. (1991). Untitled. Bethesda, Maryland: Arlen Communications Inc, March 1991.

Baer, W., and Greenberger, M. (1987). Consumer Electonic Publishing in the Competitive Environment, **Journal of Communication**, 38, pg. 49-63, 1987.

Becker, L. The Failure of Qube, in Dutton, Blumler, and Kraemer (eds.) **Wired Cities: Shaping the Future of Communications**, Boston: Hall, 1987.

Case, D. **Redefining Videotex: The Social Construction of Information Services for the Mass Audience**, unpublished manuscript, University of California at Los Angeles, 1990.

Detroit Free Press. Online or Off Target, August 12, 1991

de Fontenay, A., and Savin, B. Information Services in the U.S., presented to the International Telecommunications Society, Venice, Italy, 1990., 1990

Information and Interactive Services Report, April 26, 1991.

Information and Interactive Services Report, May 24, 1991.

Information and Interactive Services Report, June 7, 1991.

Information and Interactive Services Report, August 2, 1991.

Information and Interactive Services Report, August 30, 1991.

Jackson, C. LEC Gateways: Provision of Audio, Video and Text Services in the U.S., presented to the International Telecommunications Society, Venice, Italy, 1990.

Kramer, R. Misapplying the Minitel Model to U.S. Videotex: The Economics of Information Service Provision, presented to the International Communication Association, Chicago, 1991.

Mansell, R. **New Telecommunications Services: Videotex Development Strategies**, Information, Computer, Communications Policy Report No. 16, Paris: OECD, 1988.

Matos, F. Information Services, in **NTIA Telecom 2000 Report**, Washington, D.C.: NTIA, 1988.

Steinfield, C. and Caby, L. Strategic Organizational Applications of Videotex Among Varying Network Infrastructures, presented to the International Telecommunications Society, Venice, Italy, 1990.

U.S. Supreme Court. Feist Publications, Inc. vs. Rural Telephone Service Co., case 89-1909, Washington, D.C., 1991.

CHAPTER 13

VIDEOTEX IN A BROADER PERSPECTIVE: FROM FAILURE TO FUTURE MEDIUM?

Harry Bouwman
Department of Communication
University of Amsterdam

Mads Christoffersen
Institute of Social Sciences
Technical University of Denmark

Tomas Ohlin
Teleguide
Sweden

Many different observations can stimulate the national analyses made in this book. But considering the rich variety of specific national conditions and contingencies influencing the outcome of videotex development in each of the 13 countries, one must be extremely cautious about hasty generalizations.

However, we shall put forward several propositions on the nature of videotex as an example of a new service based on the telecommunication network, on videotex as an interlocked innovation, on marketing strategies for videotex, on the positioning of videotex in the media mix and on the policies related to introducing and further developing videotex within the European Community. These propositions are intended as a basis for further research and discussion within media, telecommunications, policy-making and market and innovation research settings.

1. The concept of videotex

First, the present experiences with videotex should be interpreted within a **historical context**. This view implies that this new communication tool must be understood as a dynamic and changing entity - not as a fixed innovation with characteristics defined once for all, but as an **innovation in process**. This development is reflected in the changes in the concept of 'videotex' itself, causing much confusion about the actual meaning of the term.

Videotex was originally a new telematic service (combining television, telephone and computer power) that was supposed to surmount the shortcomings of professional databases, which are still too specialized, too expensive and too complicated for the average consumer. Videotex was intended to be quite the opposite:
* generally accessible,
* simple to use and
* inexpensive to acquire and use.

Hence the simplistic idea of merging the two ubiquitous tools of modern communication: the telephone and the television.

The attempts to realize these aims led the constructors of the first generation of videotex to design systems with two general characteristics: graphic- and page-oriented presentation standards with 40 characters on each line and simple, standardized user interfaces. Because of national industrial policies and international competition, five different standards or protocols emerged worldwide; three, Prestel, Télétel and Bildschirmtext (BXT), were adopted in different countries in Europe. Until the mid 1980s, the question of the standards almost caused holy war between the ardent advocates. As the different standards were often supported by dedicated terminals, videotex in Europe seemed to be split up into segregated areas unable to communicate with each other.

165

H. Bouwman and M. Christoffersen (eds.), Relaunching Videotex, 165–176.
© 1992 *Kluwer Academic Publishers.*

But in recent years the issue of standards has been gradually defused, as many of the national systems have adopted different multi-standard strategies and gateways have been established to connect the systems of the 1980s.

On the terminal side, the rapid dissemination of personal computers (PCs) into private households in Europe and especially in the United States has profoundly changed the conditions for disseminating videotex services. PCs are much more flexible than dedicated terminals, and software developments are facilitating the shifts from one standard to an other. Another important trend seems to be the steady development of ASCII-based services that are targeting the private consumer market. In the United States, such pioneer services as CompuServe, GEnie and the tens of thousands of electronic bulletin board services (BBS) offer innumerable services, from software download to electronic mail and teleshopping, targeting private consumers. In the US these services are also considered to be videotex just as much as the graphic-oriented services, such as Prodigy (see Chapter 12).

The overall trend seems to be that the conception of videotex as a specific presentation standard is shifting towards a more loosely defined understanding of the medium characterized by general availability of such interactive services that put only low constraints on technology (terminals and transmission) and by the high degree of user-friendliness.

From a historical perspective videotex could be interpreted as a new medium in a **specific developmental phase** characterized by the current state of technology, politics and culture. However, this changes over time. The development of new generations of videotex systems allows more refined and flexible architectures: the advances in and declining prices of transmission technologies enable quicker build up of pages and shorter waiting times; and the advance of terminal equipment (PCs) makes it possible to establish a much more intelligent user interface.

2. Videotex as an interlocked innovation

A second point is that we believe that videotex must be understood as an **interlocked innovation**. As such, the specific innovation processes initiated imply a set of interdependent innovations in infrastructure, system provision, information and service provision and user demand. If only one of these four elements is insufficiently provided, it is very likely that the entire innovation chain will brake down and the introduction process will stop. This observation seems pertinent in trying to determine why videotex was not successful in the various countries in the 1980s.

A new service based on a familiar technology or infrastructure has a lower threshold of adoption and is often more attractive economically than a totally new technology. In this sense, videotex is a complex and complicated innovation because it is an innovation in both services and technology. We can explain this by comparing various new media and technologies and their specific combinations of service and technology.

Table 1. Adoption of an innovation

		TECHNOLOGY	
		OLD	NEW
SERVICE	OLD	Adoption/substitution + Regulation + (Colour-TV)	Adoption + Regulation + (HDTV)
	NEW	Adoption + Regulation - (Teletext)	Adoption - Regulation - (Videotex)

(Regulation: +: regulation is available; -: regulation is required.
Adoption: +: a positive attitude; -: adoption problematic).

The smoothest innovation process occurs when there is no need for major changes in the habits of the information users nor for the organization and regulation of the different production components. An example is the introduction of colour TV in the 1970s and 1980s: users' practice and media production were not essentially influenced by the innovation.
On the other hand, the introduction of teletext in the 1980s did presuppose a change in the habits and information application of the users and required some regulation. High-definition television (HDTV) will not produce substantially new user patterns, as the basic product, television, remains the same, but the production and transmission processes will be profoundly modified by this new type of television, which will tend to merge television, film and video production.

Videotex seems to represent the most complicated case in which an innovation implies both a whole new set of production procedures and a fundamental restructuring of the users' information, communication and transaction habits. Videotex thus tends to be a type of innovation that entails the most complex innovations and the most difficult adoption processes, requiring a complex of conditions to succeed. In most European countries these conditions have not been sufficiently established to produce the necessary and sufficient conditions that make videotex a success in terms of economics and user satisfaction.
The concept of interlocked innovation shows that market forces have difficulty in creating an appropriate balance between supply and demand. This book provides empirical evidence in the form of numerous national accounts of 'failure', 'mismanagement', 'user deception' and missing 'take-off' of service provision.
France, as well known, constitutes a notable exception. A crucial element in this success-story[80] is the central role of France Télécom (DGT) as the active coordinating agent. A similar position as the 'system integrator' (Quelch & Yip, 1985) seems to be attributed to Videotex Nederland in the Netherlands and to such consumer-oriented systems in the US as Prodigy. The emergence of this new type of agent can significantly influence the future marketing of videotex services.

---

[80] We shall not discuss further the precise character of the success of the Télétel system. It seems undeniable that the system will not break even before 1996-1998 (CommunicationsWeek, 2 September 1991) but the rate of return on such infrastrutural investment as a telecommunications network depends on political and economic assumptions. There are no eternal truths in this field.

3. Marketing strategies

In videotex introduction strategies, the usual approach is to target the adoption of videotex by the consumer. Even though there are exceptions, such as Sweden, the general focus has been on services for the mass market, that is the consumer. Failure to accomplish this in countries following the Prestel scenario (see Introduction) has been interpreted as inappropriate targeting of the services and the remedy has been to reorient services towards the business sector. This strategy did not prove very successful in such countries as the UK, Germany, the Netherlands, Austria and Denmark, and the negative experiences thus gained have led to a new reorientation of services for consumers in the light of the success in France.

Because of these historical conditions, very little attention has been given to the advantages of communication and transaction services for business organizations themselves. This neglect is regrettable, because videotex offers many efficient opportunities for the business market.

3.1   Strategic opportunities offered by videotex

The US Office of Technology Assessment's (OTA) report **Critical Connections** (1990) sheds some light on the question of the possible impact of computer-based communication. This can eventually affect the speed of economic transactions, the distance over which information is transported and the relationships and interdependencies among economic actors.
These three mechanisms can be combined with different goals that are strived towards within business organizations: efficiency, effectiveness and innovation. These goals are affected by every business activity, including operations, services, technology development, human resources management, firm infrastructure, logistics, procurement, marketing and sales.

Efficiency is improved by new or modified means of accomplishing existing tasks. Business processes can be accelerated, distances can be overcome and intermediaries in the business chain can be bypassed by using telecommunication. Such communication technology as videotex, yields more efficient business operations by reducing interaction time in the exchange of information. Videotex communication services makes it possible to improve communication between geographically remote offices of an organization or between an organization and employees in field offices. Videotex can improve services, for instance, by providing instructions on repairs and services, information on the availability of spare parts and so on.
The use of communication technology also results in greater **effectiveness** by improving data entry and rapid information transfer, which makes global management control possible and enables tailor-made responses to consumer demand. Another implication concerns the capability of providing information to different actors at the same time. One can imagine a videotex service for car owners of a specific type that offers information on the nearest garage and its opening times and services. An example in the field of human resource management might be training programmes offered as a videotex service both within an organization and for the consumer market.
Videotex and other communication technologies also lead to **innovation** in the form of better and new services, such as on-line databases with medical and financial information. Communication facilities also influence the relationships between actors in a production chain. One can imagine that, after the repair of a device, information could be directly stored in a database that immediately becomes available to engineers and designers, who can analyze recurring problems that might require action. An application in marketing and sales is the storage of marketing data by companies for their own purposes. But these data are also of interest to others and might even be sold to third parties.
There are three general ways of achieving competitive advantages by introducing videotex servi-

ces within an organization:

        * The costs of existing services can be reduced. Information and communication services can accelerate and improve the efficiency of business operations, both within an organization and between organizations as well as between organizations and consumers.

        * Markets can be expanded. By offering existing services through a videotex network or enhancing existing services with new applications, an organization can enter a new market segment and thus enlarge its market.

        * New services can be added to new or already existing market segments.

Steinfield & Caby (1990) create a topology of the strategies used in introducing videotex that is similar to the one mentioned in the OTA report. Essentially, this includes:

1) applications directed towards improving internal coordination and thereby lowering costs;

2) applications that help to differentiate otherwise standard products through some form of information or network service; and

3) applications that represent new products or services that can only be provided through a telecommunication infrastructure.

Based on a case-study approach, Steinfield & Caby (1990) found that the type of videotex application depends on the scope (internal, interorganizational or external) of the videotex services. A company without any interorganizational linkage pursued applications directed towards improving internal coordination. A second organization used the videotex network for product differentiation. The implementation of videotex positively affected productivity and users' orientation on information technology.

Videotex can be important developing new applications within and between organizations and in integrating new information technologies. The positive consequences of videotex for organizations and service providers is sorely neglected in adoption strategies. Organizations can meet strategic objectives through videotex-based services.

3.2.   Developing new markets: the critical mass of services

The benefits for the organization itself might be sufficient to ensure the successful introduction of videotex on the business and consumer markets but this is a gradual process. The starting point is no longer a **'critical mass of users'** but a **'critical mass of services'**. In this approach, the positive consequences of introducing videotex in an organization must be communicated to other organizations. The 'critical mass of services' can only be realized if enough organizations are convinced of the competitive advantages they will gain. Videotex will eventually spread from the business market to the consumer market.

An example is the introduction of a videotex service for flower shops offered by the Fleurop services in the Netherlands. Communication between flower shops has been based on telephone and fax. Introducing a videotex service is expected to improve both the efficiency and the effectiveness of the ordering process. Nevertheless, when consumers become aware of this service, they might bypass the local florist and order directly.

Another marketing strategy is aimed at progressively gaining a critical mass of users (Schneider et al. 1990). In the initial phase of such a **'sequencing'** (step-by-step) approach, one first tries to describe the potential added value of videotex for a specific market segment (consumer or business). It must be made clear that the service offered should satisfy clearly identifiable needs for information, communication or transaction services. An existing relationship between the service provider and user is necessary to achieve a suitable balance between the supply of services and the existing demand. After a critical mass is attained in a specific sector or target group (travel, insurance, agriculture or professionals) that recognize the utility of the service, one can proceed to another sector or target group. The first group might comprise the critical mass for the second. If the second group also joins, then a critical mass for a third group is reached, and so on.

Central to this approach is a sophisticated marketing strategy that identifies specific economic target sectors and groups to mobilize a specific group at the right time. For example this approach was followed, with limited success, by the Bundespost in Germany and the PTT in the Netherlands after they reoriented their marketing strategy and shifted their focus from the consumer to the business market. However, the sequencing approach is still followed in various strategies to introduce videotex.

3.3 The demand side: the critical mass of consumers

The various countries have different patterns of consumer and business demand for videotex services. These are mainly caused by differences in the media situation, that is the amount and quality of information and communication channels available.
An indicator of the 'success' of videotex services is the percentage of households 'possessing' a terminal (table 2). Although such figures are often outdated and their value can be questioned, they still indicate the degree to which videotex has been adopted in the various European countries. In classical innovation theory, success is defined as the number of users that have adopted (subscribed to) the innovation. By this criterion France has had the greatest success[81] followed by Switzerland, Spain and the Netherlands. However, the strongest growth in installed terminals and subscriptions is in Portugal, Spain, Italy and Switzerland.

---

[81] The figures for France are over-estimated as 20-30% of the Minitel terminals distributed generate very little or no traffic at all.

Table 2. Adoption ratios of videotex for various European countries (mid-1991)

	Number of terminals	Number of households (millions)	% of households with a terminal
Austria	12,000	2.9	0.4
Belgium	8,500	3.6	0.2
Denmark	6,500	2.2	0.3
France	5,700,000	20.5	27.8
Germany	260,000	27.8	0.9
Italy	155,000	18.6	0.8
Ireland	350	1.1	0.03
Netherlands	125,000	6.1	2.1
Portugal	4,500	3.5	0.1
Spain	325,000	11.9	2.7
Sweden	30,000	3.5	0.8
Switzerland	75,000	2.3	3.3
United Kingdom	100,000	22.0	0.5

Sources: Videotex International (1991), no 138/139, Bruno et al. (1991), and contributors to this book.

Another indicator of success is the number of connections per month and the connect time per terminal each month. Videotex is used most intensively in Germany. In Italy and Switzerland use is also high, while the Netherlands and Spain are lagging behind.

Prestel in the United Kingdom **initially** seemed to be a breakthrough in consumer telematics (1980-1984). This example was followed by Germany and many other countries, although without any success. The French Télétel system is the most used so far, and many countries consider the system as an example. The French example has led to several initiatives to (re)launch videotex in such countries as Sweden (Teleguide), the Netherlands (Videotex Nederland), Denmark (NetPlus/Info 24), Ireland and Belgium. It is interesting that Intelmatique, the international marketing branch of France Télécom, is involved in several of these initiatives.

Table 3: Number of connections and the connect time per terminal for various European countries (mid-1991)

	Number of connections by month	Connect time per terminal per month (minutes)
Austria	n.a.	30
Belgium	195,000	169
Denmark	55,000	54
France	75,000,000	100
Germany	6,248,000	455
Greece	n.a.	n.a.
Italy	1,150,000	285
Ireland	n.a.	35
Netherlands	453,000	24
Portugal	23,000	137
Spain	750,000	30
Sweden	31,000	132
Switzerland	1,104,000	205
United Kingdom	n.a.	n.a.

Sources: Videotex International (1991) no 138/139, Bruno et al. (1991) and contributors to this book.

However, we must still be very careful with this interpretation. Except for France, where videotex reaches households in general (although even this is sometimes questioned (Arnal & Jouët, 1989), videotex services in Europe are mainly business oriented. Videotex has so far failed to reach its potential market as a consumer commodity. The critical mass of users has yet to be attained in most European countries.

An important explanation for the limited adoption of videotex services can be sought in the lack of marketing research or the limited co-marketing efforts of the PTOs and the service suppliers. Underlying the videotex introduction strategies is the idea that the needs of consumers for information, communication and transactions can be satisfied by the types of services offered. However, the system operators and the service providers hardly understand the specific needs of most people and their ways of satisfying these needs. There is a fundamental lack of knowledge of the reasons why people use information technology. Information systems for the home need to be designed on the basis of a thorough understanding of the use of existing information and communication channels (Dervin, 1989).

The usefulness of videotex services for the consumer must be the central focus. User demand will be oriented towards applications. Communication and transaction services offer advantages because of their interactive nature. These types of services can be assumed to contribute to the success of videotex systems. One of the illuminating features of the French experience of the mid 1980s was the steep rise in the use of such communicative services as the famous 'messagerie roses'. These services met a need for emotional and game-like behaviour, which was possible on the Télétel network because of its open architecture and the anonymous presence of users on the kiosque services. The enormous rise in the demand for playing and emotional behaviour took completely by surprise the French planners and technocrats, who had conceived the network for utilitarian purposes. (Marchand, 1987, Charon, 1987).

Information services prove to be less important. Most of the information offered is also available through paper-based media outlets. The type of information for a consumer market does not really fulfil a fundamental need for information, contrary the situation in the professional market.

4. Positioning videotex in the media mix

The swift development of information and communication technologies has not made life simpler for the service and information providers, the telecom operators or for the users. One of the complicating factors is the abundance of different new solutions that have become available in recent years. If videotex is to survive in the long term, it must adapt to a 'media mix' of steadily growing complexity. It also must develop a profile of utility that differentiates it from the related technologies such as teletext, audiotex, compact disk read-only memory (CD-ROM), and E-mail. The challenge of these alternatives raises the following perspectives:

* Broadcast **teletext** has achieved a considerable position with the spreading of TV-sets equipped with decoders in recent years. But teletext as such has a rather narrow capacity as a telematic medium: It has limited information storage capacity and it cannot support communication and transaction services. Teletext is, however, quite efficient in disseminating up-to-date information, it has extensive coverage and is extremely competitive in user cost.

* **Audiotex** is a new medium that appears to be spreading quickly to exactly the user forum for which videotex was originally meant: the consumer (home user) or the small business user. Whereas videotex requires a visual display unit, audiotex is directly accessible by means of a simple (DTMF) phone. This makes audiotex services considerably easier to access for the inexperienced computer user that does not care about the terminal options. It is thus ironic that audiotex information provision (organized as premium rate services[82]) is growing quickly at present just as was forecasted for videotex ten years ago.

The advantages of audiotex in terms of ubiquity and user-friendliness, however, are balanced by its disadvantages compared with videotex. Many sophisticated services depend on more complex input than can be provided by the 12 buttons on the DTMF phone: they require an alpha-numeric keyboard. Another important factor is the limited flexibility of audiotex in searching for and retrieving information. As the information output is only voice, the user has to memorize the alternatives; most people cannot simultaneously keep more than three or four items in mind.
Until audiotex has been developed further to master speech recognition as an efficient means of information reception, most voice- or sound-based systems will have limited capability for the more sophisticated transaction services. Even if audiotex occupies a prominent position in the private comsumer market, it will probably not completely "cannibalize" videotex services. These will find their niche where the complexity of search procedures and of the information output exceeds the capabilities of the audiotex systems. Nevertheless, many 'low key' information systems can easily find an efficient user profile with voice-based systems, and this implies a reduction of the total telematic market open to videotex.

* **CD-ROMs** (and other compact disks) are expected to be increasingly propagated because they have a vast capacity to store information in the form of text, data, graphics and sound. But CDs demand relatively costly equipment (PC, CD drive and special software) and their dissemination

---

[82] 0898-services in the UK, 900-services in US and Denmark, 071-services in Sweden, and 06-services in the Netherlands.

so far has been limited to professional settings. The advantages of CD-ROM compared with videotex are that there is no transmission cost once the system is in operation and the information can be reused perpetually. Nevertheless, with the currently available CD-ROMs, the information can only be updated by distributing new disks.

* **E-mail** was expected to grow substantially in the 1980s but these optimistic hopes were dashed. Fax transmission has been booming, whereas electronic mail has only become popular in settings with a highly developed computer culture. Public subscription to advanced E-mail services in accordance with the X.400 standard has been marginal in most countries. In France the Télétel system has been augmented with an E-mail service (Minicom) but the traffic has been quite moderate[83].

Electronic mail may eventually be incorporated in future videotex systems in Europe: this is the case in the United States, where such services as CompuServe and Prodigy have mail systems that are used extensively. But these functionalities largely depend on the level of sophistication of the terminal, and it is difficult to imagine successful E-mail for private users and small businesses, that is not based on the 'intelligence' of the PC.

5. Videotex in the 1990s: policy aspects

The development of videotex in the 1970s and 1980s was driven by technology and policy. The early Prestel scenario was especially pushed by engineers from the PTTs in the various countries, whereas the dominant motivation for the Télétel operation seems to have been the anxiety of French Government and technocracy about lagging behind in the technological race with Japan and the United States. Both scenarios express policies deriving from the monopolistic regulatory framework with only few limitations for the PTTs. This situation enables a massive transfer of resources from one service area to another: cross-subsidization. Télétel is the most famous example of this, as the distribution of 'free' terminals for videotex was paid for by the income of France Télécom in other branches, especially traditional telephony.

The gradual introduction of competition within telecommunications has placed narrow constraints on the companies' possibilities of practicing cross-subsidization. The Commission of the European Communities with its General Directorate XIII as a driving force has played a dominant role in imposing these important reorganizations.
The deregulation process will significantly affect the development of videotex services in Europe. Videotex, like other services based on telecommunication network, will no longer benefit from subsidies from other sources, and the operation of networks will be open to competition - similar to the provision of services. These changes have already influenced the behaviour of the operating telecommunication companies. As they no longer (or only for a short period) have a monopoly on network operation, they are constrained to orient their behaviour towards the competitive market. It can be considered as adaptation to these conditions when PTTs such as those in the Netherlands and in Sweden engage in joint ventures with other partners to set up such specialised videotex companies as Videotex Nederland and Teleguide.
Another consequence is the more or less open reluctance of PTTs to invest in developing infrastructure for the service provision because they fear losing more money on an already unprofitable system: for example in Belgium and Denmark (see chapters 7 and 8).

---

[83] The Minicom received around 300,000 calls per month in first four months of 1991 which represents 0.05 % of the amount of calls to the electronic directory (60 million per month)(La lettre Télétel, no 22, 1991).

The European Commission tried to play an active role in promoting videotex services in Europe. In 1990 the Commission persuaded representatives of videotex services in 17 countries in Europe to sign an agreement to connect the various videotex systems. This interconnection makes access according to the different standards (Prestel, BTX, Télétel and ASCII) possible. At the management level this interconnection is a fact; the technical implementation is progressing whereas there are substantial problems in setting up a flexible invoicing procedure. In most cases is it not possible to gain direct access to another national system without a separate subscription.

The future development of videotex in Europe depends largely on the ways in which market forces and public policy-making will react to the new and changing conditions. The scene is still open to new actors but the risks of participating are considerable. The Prestel scenario with its centralized architecture has been no success in the 1980s. The Télétel operation has been a relative success in France but it still has to prove that it is viable outside the very specific French conditions. The question is also if whether or not the lack of intelligence in the Minitel will block further development of advanced services such as E-mail, teleshopping and banking.

From the studies presented in this book we can learn that a multitude of variables are influencing the final outcome: life or death for videotex. One of these variables concerns 'system integration'. Videotex can only be successful if it is coordinated by some integrating organization: such as the PTTs in France and Italy, the banks in Switzerland or big corporations in the United Stated (Prodigy).

Another group of variables relates to marketing. In the case of videotex, which tends to be technology-driven, marketing requires fundamental research in the information and communication behaviour of consumers, professionals and business users. The knowledge of these mechanisms has been largely insufficient.

A third group of variables is related to the quickly changing media environment of the 1990s. Videotex will without doubt be exposed to competition from other telecommunications network-based services.

The developing of new services and new technologies is clearly a learning process including a multitude of actors, that influence the final result in complex ways. For those engaged in videotex the sentiments have varied from hope and optimism to desperation and tidings of death. Time will show if there is a life after death for videotex.

References.

Arnal, N. & Jouët, J. (1989). Télétel: images des utilisateurs résidentiels. **Technology de l'Information et Société, Vol 2 (1) & Résaux (37),** pp. 105-125.

Bruno, S., Cohendet, P., Desmartin, F., Llerena, D. , Llerena, P. & Sorge, A. (1991). **Modes of Usage and Diffusion of New Technologies and New Knowledge. A Synthesis Report.** Brussel: Commision of the European Communities, Monitor, FAST programme.

Charon, J-M (1987). Télétel, de l'interactivité homme/mashine à la communication médiatisée. In Marchand (ed.). **Les paradis informationnels**. Paris.

Dervin, B. (1989). Users as research Inventions: How research categories Perpetuate Inequities. **Journal of Communications.** 19 (3) 216-232.

OTA (1990). **Critical connections**.

Marchand, M (1987). **La grande aventure du Minitel**. Paris.

Quelch, J.A. & G.S. Yip (1985). Achieving System Cooperation in Developing the Market for Consumer Videotex. In: R.D. Buzzell (ed). **Marketing in an Electronic Age.** Boston.

Steinfield, C. & Caby, L. (1990). **Strategic applications of videotex in user organizations among varying network infrastructures**. Presented to the international Telecommunications Society. Venice, March.

Williams, F., Rice, R. & Rogers, E. (1989). **Research Methods and the New Media**. New York: the Free Press.

INDEX